COAST-TO-COAST AC

OUR STOLEN

THEO COLBORN is a senior scientist with the World Wildlife Fund
and a recognized expert on endocrine-disrupting chemicals. She
received her Ph.D. in zoology from the University of Wisconsin at
Madison and speaks regularly to scientific groups, health officials, and
policy makers. She lives and works in Washington, D.C. DIANNE
DUMANOSKI has reported on national and global environmental is-
sues for the *Boston Globe* and was the recipient of the prestigious
Knight Fellowship in Science Journalism at MIT. She lives outside
Boston. JOHN PETERSON MYERS is director of the W. Alton Jones
Foundation, a private foundation supporting efforts to protect the
global environment and to prevent nuclear war; he was formerly
Senior Vice President for Science at the National Audubon Society.
He received his Ph.D. in zoology from the University of California,
Berkeley, and he lives near Charlottesville, Virginia.

"Could become the biggest scientific and public-relations bombshell to hit the chemical industry since Rachel L. Carson's 1962 classic, *Silent Spring*."
—*Business Week*

"Meticulous, readable . . . pulls together an astounding number of research findings."
—*Los Angeles Times*

"The first book to weave decades of research on hormone disruptors into a single, disturbing picture."
—*USA Today*

"Reads like a real-life scientific thriller."
—*Texas Observer*

"Just as Rachel Carson's writings stirred the public of the '60s, this new book should bring on a revolt against chemical pollution."
—*San Francisco Examiner-Chronicle*

"A strong hypothesis that raises troublesome questions about chemicals that are in almost every home . . . promotes a radically different way of viewing chemical hazards."
—*Seattle Post-Intelligencer*

"The economic stakes involved are so huge that the regulatory quandaries we could soon be mired in will make earlier struggles look like kids' stuff."
—*Cleveland Plain Dealer*

"A significant book about a major environmental issue . . . catalogues and pulls together a vast amount of information collected by wildlife biologists."
—*Providence Sunday Journal*

"A compelling book on a serious scientific/environmental societal issue, disturbing in its implications for the future of humankind."
—*Buffalo News*

"An important and successful attempt to warn the public of a possibly catastrophic technological danger. . . . Anyone who cares about the health and even survival of future generations should read it."
—*Roanoke Times*

OUR STOLEN FUTURE

*Are We Threatening
Our Fertility, Intelligence,
and Survival?—A Scientific
Detective Story*

WITH A NEW EPILOGUE
BY THE AUTHORS

Theo Colborn,
Dianne Dumanoski,
and John Peterson Myers

A PLUME BOOK

To all of you who contributed to this effort, each in your own way—with your expertise, your critical assessment, your time, your encouragement, and your patience—and for your children and grandchildren and their grandchildren.

PLUME
Published by the Penguin Group
Penguin Books USA Inc., 375 Hudson Street, New York, New York 10014, U.S.A.
Penguin Books Ltd, 27 Wrights Lane, London W8 5TZ, England
Penguin Books Australia Ltd, Ringwood, Victoria, Australia
Penguin Books Canada Ltd, 10 Alcorn Avenue, Toronto, Ontario, Canada M4V 3B2
Penguin Books (N.Z.) Ltd, 182–190 Wairau Road, Auckland 10, New Zealand

Penguin Books Ltd, Registered Offices: Harmondsworth, Middlesex, England

Published by Plume, an imprint of Dutton Signet,
a division of Penguin Books USA Inc.
Previously published in a Dutton edition.

First Plume Printing, March, 1997
10 9 8 7 6 5 4

 REGISTERED TRADEMARK—MARCA REGISTRADA

The Library of Congress has catalogued the Dutton edition as follows:
Colborn, Theo.
Our stolen future : are we threatening our fertility, intelligence, and survival? : a scientific
detective story / Theo Colborn, Dianne Dumanoski, and John Peterson Myers.
p.   cm.
ISBN 0-525-93982-2 (hc.)
ISBN 0-452-27414-1 (pbk.)
1. Reproductive toxicology.   2. Environmental health.   I. Myers, John Peterson.
II. Dumanoski, Dianne.   III. Title.
RA1224.2.C65   1996
615.9'02—dc20   95–30015
CIP

Printed in the United States of America
Original hardcover design by Jesse Cohen

FOREWORD

Vice President Al Gore
January 22, 1996

Last year I wrote a foreword to the thirtieth anniversary edition of Rachel Carson's classic work, *Silent Spring*. Little did I realize that I would so soon be writing a foreword to a book that is in many respects its sequel.

Thanks to Rachel Carson's clarion call, we developed new and vital protections for the American public. Now *Our Stolen Future* raises questions just as profound as those Carson raised thirty years ago—questions for which we must seek answers.

Silent Spring was an eloquent and urgent warning about the dangers posed by manmade pesticides. Carson not only described how persistent chemicals were contaminating the natural world, she documented how those chemicals were accumulating in our bodies. Since then, studies of human breast milk and body fat have confirmed the extent of our exposure. Human beings in such remote locations as Canada's far northern Baffin Island now carry traces of persistent synthetic chemicals in their bodies, including such notorious compounds as PCBs, DDT, and dioxin. Even worse, in the womb and through breast milk, mothers pass this chemical legacy on to the next generation.

As Carson warned in one of her last speeches, this contamination has been an unprecedented experiment: "We are subjecting whole populations to exposure to chemicals which animal experiments have proved to be extremely poisonous and in many cases cumulative in their effects. These exposures now begin at or before birth and—unless we change our methods—will continue through the lifetime of those now living. No one knows what the results will be because we have no previous experience to guide us."

We are only now beginning to understand the consequences of this contamination. *Our Stolen Future* takes up where Carson left off and reviews a large and growing body of scientific evidence linking synthetic chemicals to aberrant sexual development and behavioral and reproductive problems. Although much of the evidence these scientific studies review is for animal populations and ecological effects, there are important implications for human health as well.

A decade ago, the ozone hole provided shocking evidence of the atmospheric effects of chlorofluorocarbons (CFCs). Last year, scientists declared that human activity is changing the earth's climate. Today, reports in leading medical journals point ominously to hormone-disrupting chemicals' effects on our fertility—on our children.

Our Stolen Future provides a vivid and readable account of emerging scientific research about how a wide range of manmade chemicals disrupt delicate hormone systems. These systems play a critical role in processes ranging from human sexual development to behavior, intelligence, and the functioning of the immune system.

Although scientists are just beginning to explore the implications of this research, initial animal and human studies link these chemicals to myriad effects, including low sperm counts; infertility; genital deformities; hormonally triggered human cancers, such as those of the breast and prostate gland; neurological disorders in children, such as hyperactivity and deficits in attention; and developmental and reproductive problems in wildlife.

The scientific case is still emerging, and our understanding of the nature and magnitude of this threat is bound to evolve as research advances. Moreover, because industrial chemicals have become a major sector of the global economy, any evidence linking them to serious ecological and human health problems is bound to generate controversy. However, it is clear that the body of scientific research underlying *Our Stolen Future* raises compelling and urgent questions that must be addressed.

Responding to the mounting evidence, the National Academy of Sciences has established an expert panel to assess the threats. That is an important first step. We must also expand research efforts to learn more about how these chemicals may do their damage, to identify how many other synthetic chemicals possess such properties, and to discover the extent to which we and our children are exposed. We need to understand the often invisible damage they may cause. We must find out if there are ways to protect children, who appear to be at greatest risk for birth defects and developmental disorders from such hormonally active compounds. We need to explore further the links between effects on humans and those on wildlife.

We can never construct a society that is completely free of risk. At a minimum, however, the American people have a right to know the substances to which they and their children are being exposed and to know everything that science can tell us about the hazards.

It is now clear that we waited too long to ask the right questions about the CFCs that eventually attacked the ozone layer, and we are going too slow in addressing the threat of climate change. We certainly waited too long to ask the right questions about PCBs, DDT, and other chemicals, now banned, that presented serious human health risks.

Our Stolen Future is a critically important book that forces us to ask new questions about the synthetic chemicals that we have spread across this Earth. For the sake of our children and grandchildren, we must urgently seek the answers. All of us have the right to know and an obligation to learn.

ACKNOWLEDGMENTS

This book was a collaborative effort that extended well beyond us to include scientists, scholars, and friends from around the world. It would be impossible to name everyone who gave their time and expertise in the development of this story. Rather than risk the chance of omitting anyone, we have chosen not to produce a list of those who deserve thanks. We hope that each one of you (you know who you are), as you read the book, will be able to take satisfaction in what we have produced. We are truly indebted to you.

In addition, this book would never have reached publication without the initial and constant support of a number of foundations: the W. Alton Jones Foundation, the Joyce Foundation, the C. S. Mott Foundation, the Pew Scholars Program, Pew Charitable Trusts, the Winslow Foundation, and the Keland Endowment of the Johnson Foundation.

CONTENTS

PROLOGUE

This is a most unusual book. The product of a collaboration by three authors, it uses an untraditional style to present a message that transcends traditional knowledge concerning synthetic chemicals, their safety, and how we perceive risk. The three of us who worked on this book—Theo Colborn, Dianne Dumanoski, and Pete Myers—each brought different talents and experience to the task and played a different role in getting this book to press. We entered into this collaboration because the increasingly complex problems facing us at the close of the twentieth century demand such cooperative efforts. They require more than any single individual can bring to the challenge.

Theo Colborn's seven years of work synthesizing the research on endocrine-disrupting chemicals and her extensive data base provided the scientific foundation for this effort. Dianne Dumanoski's challenge was to take the complex science and transform it into a story that would be accessible to everyone, including those without any scientific background. Dianne, who has reported and written about environmental science and policy for twenty-five years, supplemented this information through additional research and interviews. Pete Myers brought a background in science, as well as extensive experience in national and international environmental policy, adding another valuable dimension to our thinking. The authors developed and refined the book's structure and argument together, working closely and regularly in long sessions through much of the writing.

Since this is still an unfolding scientific mystery, it is told as a detective story with Theo Colborn and Pete Myers appearing as figures in the text, as do the other scientists who have played key roles. The first part of this story takes the reader through Theo's process of discovery as she reviewed the scientific literature concerning the health effects of synthetic chemicals on wildlife and humans. Theo is the sleuth in this scientific mystery not only because she really played such a role but also because we think this approach will engage the reader. As the book moves beyond Theo's early detective work, it begins to discuss the evidence and reflects the thinking of all three of us.

We live in a complex world that is going to require innovative approaches to deal with the problems technology has created. It has taken a nontraditional approach—involving extensive cooperation among experts from many disciplines—to reveal the nature of the chemicals that are stealing our future. Just as scientists had to break with convention to uncover this problem, we found we had to break with literary convention to tell the story of their discovery.

PREFACE TO
PAPERBACK EDITION

In the eight months since its publication in March 1996, *Our Stolen Future* has accomplished much of what we had hoped when we set out to write a popular book about the disturbing new scientific evidence that many synthetic chemicals can disrupt hormones. We aimed to take this urgent message to a broad audience beyond the scientific community. Ours, we knew, was far from the first warning about hormone-like effects in some synthetic chemicals. As early as 1950 researchers reported that the man-made pesticide DDT could derail the sexual development of roosters and, through what appeared to be hormonal action, lead to "chemical castration." But that scientific report sank into oblivion, the warning went unheeded, and for decades vast quantities of such chemicals continued to be spread across the face of the earth. Fortunately, our own effort has not suffered the same fate.

The release of *Our Stolen Future* sparked a tremendous amount of discussion and controversy and a blizzard of media coverage, not only in the U.S. but in other countries as well. Thanks to this intense attention, the idea that synthetic chemicals pose significant health risks by disrupting hormones has taken firm root in the public debate, and the issue of endocrine disruption, as scientists call it, seems to have won a place on the public policy agenda. As hoped, the book has succeeded in challenging the prevailing notion that the primary hazard posed by toxic chemicals is cancer. The increasing

prominence of the theory laid out in *Our Stolen Future* has stimulated a flurry of scientific research and dozens of scientific panels,
meetings, and conferences. The U.S. Congress has responded with
four pieces of legislation this year that mandate various actions including new procedures to screen chemicals for hormonal effects.
The Clinton administration announced a new policy to protect children's health from environmental threats, including hazards from
chemicals that disrupt hormones. According to EPA administrator
Carol Browner, the initiative aims to ensure that health standards
and risk calculations reflect the special vulnerability of infants and
children. Through translations and twelve foreign editions to date,
the book is reaching audiences around the world and contributing to
the gathering momentum for a global treaty to control and eliminate
persistent chemicals, including hormone-disrupting compounds
such as DDT and PCBs.

All of this gives us great hope. The scientific evidence laid out
in this book raises questions that urgently demand answers and action. The recognition that synthetic chemicals can disrupt hormone
messages in the body may well have implications as profound as the
discovery that man-made chlorofluorocarbons (CFCs) can attack
the Earth's protective ozone layer. In our view, the evidence clearly
indicates that these chemicals are already affecting some humans as
well as many species of wildlife, but at the moment the magnitude
and seriousness of this problem are still unknown. Numerous studies, however, show that astoundingly small quantities of these hormonally active compounds can wreak all manner of biological havoc,
particularly in those exposed in the womb. During this creative and
vulnerable time, the chemical messages carried by hormones trigger
key events essential for normal development, ranging from the sexual
differentiation of males to the orderly migration of nerve cells necessary for the construction of a well-functioning brain. By scrambling
these hormone messages, synthetic chemicals can undermine a baby's
development with consequences that will last a lifetime. Wildlife
studies vividly demonstrate that these chemicals have the power
to derail sexual development, creating intersex individuals that are
neither male nor female. They can sabotage fertility, erode intelligence, undermine the immune system, and alter behavior.

In the months since this book was completed, a number of important new studies have appeared that add to our knowledge of such controversial questions as trends in human sperm count and lend further weight to concerns that such chemicals are already taking a toll on human health and well-being. We will review the latest scientific developments in the epilogue at the end of this edition.

Even before *Our Stolen Future* first hit the bookstores, a noisy publicity campaign was already underway to portray these concerns as unfounded alarmism. As anticipated, some immediately charged that this book exaggerates the possible dangers and amounts to another "chicken little" scare about environmental hazards. A few conservative commentators and publications even resorted to ad hominem attacks in editorials and op-ed pieces impugning our scientific credibility and professional integrity. None of this was surprising. In fact, given the vicious attacks on Rachel Carson following the release of *Silent Spring* in 1962, we had steeled ourselves for far worse. Many of these attacks featured flamboyant rhetoric coupled with a notable lack of scientific substance. Most were inspired by a handful of ultra-conservative advocates who receive industry funding for their efforts to campaign against laws that protect health and the environment.

In marked contrast to its concerted campaign thirty-four years ago to discredit Rachel Carson and her message, however, the chemical industry itself has been remarkably low-key and cautious in its public response to *Our Stolen Future*. The industry associations and individual companies have refrained from direct attacks on the book and have issued uncontentious public statements acknowledging that the book raises serious questions. Some have sought to downplay concerns by claiming that their chemical testing programs have been screening for endocrine and offspring effects all along—a claim disputed by the Environmental Protection Agency and many scientists cited in this book.

At the same time, the book has not gone unchallenged. Some commentators have identified legitimate scientific issues that require investigation. Nor do we dispute the need for additional research on many of the questions we raise about the impacts of low-level contamination on human health. Nevertheless, we are convinced enough is known already to justify taking the steps recommended in Chapter 13, and we are gratified that some of these initiatives are already underway.

Other challenges, however, were bent on broad rejection of the overall thesis. Some industry-supported scientists and academics devoted a remarkable amount of time and energy to dismissing questions about possible links between human health problems and chemical contamination. The arguments raised in this regard often showed little familiarity with the book or the underlying science. Sometimes they rested on an inappropriate use of data that would be obvious to anyone with scientific training, raising the question of whether the object was earnest scientific debate or propaganda aimed at the media and the public.

Just as it is difficult, with current knowledge, to affirm *all* the human health impacts we raise in the book, it is impossible to dismiss biologically plausible theories that hormone-disrupting chemicals may be a factor eroding human health, undermining intelligence, and reducing reproductive capacity.

The chemical industry both directly and through its trade associations is also funding its own research into the endocrine-disruption problem. This is a step in the right direction, but unfortunately, past experience suggests that this money will most likely flow to researchers inclined toward the industry view and yield studies to counter evidence of possible links between human health problems and synthetic chemicals. While skepticism is healthy and theories such as the endocrine-disruption hypothesis need to be challenged and tested, it would be better to have disinterested scientists do such testing with funds that are not tied directly to industry. Few would argue that industry-funded science is automatically or necessarily bad science. But studies funded by interests that have millions of dollars at stake in the outcome suffer at the outset from a credibility problem and only serve to further polarize the political debate. Individual studies may be well executed and unbiased, but the answers they yield may depend to a large extent on study design and how the question is posed. We regret that this money has not gone into some independent fund to support carefully designed and coordinated studies overseen by a disinterested panel of leading scientists.

The signs at the moment suggest that vested interests will focus their efforts on the Environmental Protection Agency's process to develop new screening procedures for identifying endocrine disruptors. The struggle will center on how broadly to define the problem. Factions within the chemical industry have already attempted to limit the scope of concern solely to compounds that interfere with estrogen. As the weight of the evidence has shown, the issue is far broader. Chemical interests have much to gain from denying this process and restricting the official scope of the endocrine-disruptor problem as much as possible. The narrower this definition, the easier it will be to confine the problem and dismiss its importance, and the cheaper it will be to conduct new tests.

The most remarkable thing about the media response to *Our Stolen Future* has been the sheer quantity and breadth of the coverage. In recent months, hundreds of news stories, book reviews, radio and TV interviews, magazine articles, and talk-show programs have focused on this book and this important issue. Through publications ranging from the journal *Science* to *Esquire* and *Garden Design*, news of our concerns has managed to reach a wide variety of audiences.

News accounts in the science press and even in industry trade magazines have, by and large, given thoughtful coverage to the book, neither dismissing us as alarmist nor disputing the fact that this scientific case demands serious consideration. One notable exception was a story in the science section of one leading newspaper which undertook to portray the book's concerns as baseless and to debunk the warnings it raises.

It may be inevitable that complex issues suffer some degree of caricature in press coverage and public debate. Reports typically focused only on estrogenic chemicals, neglecting equally compelling evidence that synthetic chemicals can block testosterone or disrupt the thyroid hormones necessary for brain development. The discovery of interference with other hormones is a critical part of this unfolding story, since it demonstrates that synthetic chemical disruption involving estrogen isn't a unique case, a fluke stemming from some peculiarity with the estrogen receptor. There are many routes to hormone havoc, and they are clearly not limited to estrogen or to a biological effect involving a hormone receptor. The emerging science, in fact, demonstrates that much of the body's chemical messaging system—not just the endocrine system—is vulnerable to disruption by the novel compounds that humans have synthesized and released into the environment.

Perhaps unavoidably, journalists also tended to seize on a few human health problems, such as the controversy over reports of falling sperm counts or possible environmental links to rising rates of breast cancer. This approach was immensely frustrating because it was so contrary to the method of this book and the nature of its argument. In our scientific detective story we lay out an extensive body of evidence drawn from many different fields and types of scientific work—wildlife studies, lab experiments, mechanistic investigations of hormone receptors, reports on contamination levels in humans and wildlife, and human data from the tragic medical misadventure with the synthetic estrogen DES. The questions about possible effects on human health, which we finally consider in Chapter 9, arise from what the totality of the biological evidence shows about the damage done by hormone-disrupting chemicals, human exposure to these contaminants, and potential human vulnerability. The wildlife studies and lab experiments suggest the possibility of increasing rates of testicular cancer, male genital defects, and suppressed sperm counts in humans exposed to synthetic chemicals that mimic estrogen. If one does find scattered reports of increasing genital defects, these prompt more concern when considered in the light of this body of supporting animal evidence. If one detects worrisome trends in a variety of health problems linked to hormone disruption, the pattern may be far more telling than *any* single trend. It is a profound mistake to view this issue through the narrow lens of a single illness or physical deficiency or to consider human evidence alone. The power of this book's argument rests on the cumulative weight of the evidence and the compelling patterns that emerge when it is considered as a whole.   —*November 1996*

Since this story is far from over, we will continue our exploration of the endocrine disruption problem at a special Web site. For the latest developments in both science and policy, please join us at www.osf.facts.org

1

OMENS

1952: Gulf Coast, Florida

In YEARS OF WATCHING BALD EAGLES, CHARLES BROLEY HAD NEVER
seen anything like it, so he made careful notes in his field diary, a
record that would over time document the decline of the bird along
the eastern coast of Canada and the United States. Broley, a Canadian, made his living as a banker, but he worked with equal intensity
at his passionate avocation—ornithology. Long before the abandoned
nests with broken eggshells appeared, he noticed the eagles were acting strangely.

Broley's study of Florida's bald eagles had begun in 1939 at the
suggestion of staff at the National Audubon Society. After his initial
surveys, he made enthusiastic reports about a robust eagle population that was nesting successfully all along the peninsula's west coast
from Tampa to Fort Myers. In the early '40s, Broley followed 125 active nests and, climbing aloft, he banded some 150 young eaglets
each year.

Then in 1947, the picture suddenly changed. The number of

eaglets began dropping sharply, and in the succeeding years, Broley witnessed bizarre behavior in many of the eagle pairs. The season was early winter, the time when adult birds find mates and begin their courtship by gathering sticks and building a nest together. But at nesting sites he had visited for thirteen years, two-thirds of the adult birds, easily recognized by their white heads, appeared indifferent to the nesting ritual. They engaged in no courtship activity. As Broley noted in his diary, they showed no interest whatsoever in mating. The birds just "loafed."

What had caused Florida's eagles to lose their natural instinct to mate and raise young? When Broley looked around for a possible explanation, his eyes fell on the large housing developments spawned by Florida's postwar development boom. The new homes were gobbling up hundreds of acres of prime coastal land, so Broley attributed the declines and aberrant behavior to human intrusion. University-based eagle researchers fully concurred in this initial diagnosis.

Broley later came to doubt this explanation. As he continued his work through the mid-1950s, he became firmly convinced that 80 percent of Florida's bald eagles were sterile—a complaint that one could hardly blame on bulldozers.

The late 1950s: England

Although otters were no longer as plentiful as in earlier times, the traditional sport of otter hunting continued into midcentury relatively unchanged from the days when Sir Edwin Landseer had captured the kill in his nineteenth-century oil painting *The Otter Hunt*. Devotees of the sport in Britain still kept at least thirteen packs of shaggy, long-eared hounds for the chase and feisty little terriers to flush the otters out. And those who had learned the habits of the otter from fathers and uncles before them still knew where to seek out the otter's lair. On weekends during the hunting season, they would scout along the stream banks looking amid the tangled roots for the hollows that shelter otters by day. Once an otter was on the run, the sounds of a horn and the baying of the hounds would resonate through country glens as men pursued an ancient blood sport.

By the end of the decade, however, hunters began to have trouble finding otters to hunt, and then in some areas otters disappeared altogether for no apparent reason.

Other than the hunters, few noticed that the elusive and largely nocturnal animals were vanishing from rivers and streams where they had always been found. When conservationists finally took note of the problem almost two decades after the decline had begun, they looked to the huntsmen's records for clues about why the otters had disappeared.

Some suspected the pesticide dieldrin, but the cause of the decline would remain a mystery until the 1980s, when English scientists analyzed evidence from across Europe.

The mid-1960s: Lake Michigan

In the post–World War II economic boom, the consumer appetite for new luxuries seemed insatiable. For Michigan mink ranchers, these were good times indeed, as the wave of prosperity carried them from one flush year to another through the 1950s. Pat Nixon might be wearing a good Republican cloth coat, but other American women wanted mink.

By the early 1960s, however, the mink industry that had grown up around the Great Lakes because of the ready supply of cheap fish began to falter—not because the demand for mink was flagging but because of mysterious reproductive problems. Ranchers were mating their domesticated mink as they always had, but the females weren't producing pups. At first, the average number of pups had dropped from four to two, but by 1967, many females never gave birth, and the few that did inevitably lost those babies soon afterward. In some cases the mothers died as well. The only mink ranchers escaping the devastating losses were those who fed their animals with fish imported from the West Coast.

Michigan State University researchers, seeking to identify the cause, immediately zeroed in on the pollutants carried in Great Lakes fish, eventually linking the reproductive failure to PCBs, a family of synthetic chemicals used to insulate electrical equipment.

Curiously, other mink ranchers in the Midwest had also faced financial ruin because of reproductive failure a decade earlier. But in their case, the mink herds crashed after the animals were fed with scraps from chickens that had been given the synthetic drug diethylstilbestrol or DES, a man-made female hormone, to make them grow faster. Although the symptoms were strikingly similar, the second crash among fish-fed mink could not be linked to DES, and the connection between the two declines remained a mystery.

1970: Lake Ontario

The sight of the herring gull colony on Near Island was overwhelming, even for a seasoned biologist like Mike Gilbertson. This was the time when the gulls should be busily feeding their squawking, demanding brood, but what the Canadian Wildlife Service biologist found instead was a scene of devastation. As he walked through the barren sandy expanse where the gulls breed and raise their young, he encountered unhatched eggs and abandoned nests everywhere, and here and there, dead chicks.

In a quick count, Gilbertson estimated that eighty percent of the chicks had died before they hatched, an extraordinary number. As he examined the dead chicks, he saw grotesque deformities. Some had adult feathers instead of down, club feet, missing eyes, twisted bills. Others looked shriveled and wasted, and they still had the yolk sack attached, suggesting they hadn't been able to use its energy for development.

Something about the symptoms seemed vaguely familiar, but Gilbertson knew he had never seen them in the field. Where had he heard about this before? The question nagged at him as he completed his melancholy tour and headed back by boat to his laboratory.

A few days later, it suddenly came back to him. Chick edema disease—an affliction he had read about as a student in England. The same deformities and wasting had shown up in the offspring of chickens exposed to dioxin in laboratory experiments. If the dead gulls had all the symptoms of chick edema disease, he thought, there must be dioxin contamination in the Great Lakes.

Gilbertson's colleagues and superiors greeted this theory with skepticism bordering on derision. Some doubted his diagnosis because dioxin had never been reported in the lake, and their doubt only deepened when analysis of gull eggs with the methods then available could find no trace of dioxin.

Gilbertson nevertheless remained convinced that Great Lakes birds were showing signs of dioxin contamination, but he found no support for pursuing his theory.

The early 1970s. The Channel Islands,
Southern California

Even the trained eye finds it difficult to tell a male western gull from a female. So if it hadn't been for the extra eggs in the nests, it is possible no one would have made the startling discovery that females were nesting with other females.

Ralph Schreiber of the Los Angeles County Natural History Museum first spotted nests with unusually large numbers of eggs on San Nicolas Island in 1968. Gulls find it difficult to incubate more than three eggs at a time, so he immediately suspected that more than one female must be laying in these nests.

Four years later, George and Molly Hunt from the University of California at Irvine discovered the same phenomenon on Santa Barbara, a smaller island closer to shore. At least eleven percent of the nests on that island had four to five eggs, and they found that in these nests fewer chicks than normal actually hatched. The Hunts saw thinning eggshells in the Santa Barbara gull colony as well, leading them to suspect that the birds were suffering from exposure to DDT.

The Hunts could not at first confirm whether females were, indeed, sharing nests. In later work, however, the husband-and-wife team did establish that the female gulls were setting up housekeeping together and producing the nests with extra eggs. In a 1977 paper in the journal *Science*, they explored possible natural explanations for such behavior, suggesting that same-sex pairings might be an adaptation that conferred some evolutionary advantage.

Over the next two decades, female pairs would be found nesting in populations of herring gulls in the Great Lakes, among glaucous gulls in Puget Sound, and among the endangered roseate terns off the coast of Massachusetts.

The 1980s: Lake Apopka, Florida

Judging by the lush wetlands along its shore, Lake Apopka, one of Florida's largest water bodies, should be alligator heaven. The lake was understandably high on the list when state and federal wildlife researchers began looking for a source of eggs for the state's multi-million-dollar alligator ranching industry, which raises the reptiles for their handsome hides. To their surprise, however, the biologists found that Apopka's alligators had no eggs to spare.

In some Florida lakes, the surveys showed that ninety percent of the eggs laid by female alligators hatch. At Lake Apopka, the hatching barely reached eighteen percent. Even worse, half of the baby gators that hatched languished and died within the first ten days.

Lou Guillette, a University of Florida specialist on the reproductive biology of reptiles, could not make sense of the symptoms he was seeing. There seemed little question that the problems found in the lake's alligators were somehow linked to the 1980 accident at the Tower Chemical Company, located a quarter of a mile from the lake's shore. Right after the spill of the pesticide dicofol, more than ninety percent of the alligator population had disappeared. But why were the alligators suffering reproductive problems long after water samples suggested that the lake was again clean?

When researchers took to the water at night in airboats to trap the lake's alligators and examine them more closely, they found a strange deformity in many of the males: at least sixty percent of them had abnormally tiny penises. Nothing like this had ever been reported before.

What kind of toxic effect could this possibly be?

1988: Northern Europe

The first signs of the epidemic that would become the largest seal die-off in history appeared in the spring on the island of Anholt in the Kattegat, a narrow strait between Sweden and Denmark.

In mid-April, those conducting routine surveys of the seal populations began to find aborted harbor seal pups washed up on the wet sand along with the debris from winter storms. Not long after, the tides carried in the speckled silver bodies of older animals as well.

Because of the contamination in Europe's coastal waters, some immediately speculated that the animals were victims of pollution, but Dutch virologist and veterinarian Albert Osterhaus was skeptical from the start. All signs pointed to some infectious disease.

By the end of the month, reports of more dead seals were coming in from Hesselø, a smaller, less accessible island to the south. From there, the die-off spread at a gallop along the coastal areas throughout the North Sea, hitting seals in the Skagerrak Strait between Denmark and Norway in June, the herds in the Oslo Fjord in July, and the harbor seals on England's east coast in early August. By September, the beaches of the remote Orkney Islands off the northern tip of Scotland, the Scottish west coast, and the Irish Sea were also awash in dead seals. The death toll by December came to almost eighteen thousand seals, more than forty percent of the entire North Sea population.

Strangely, however, the plague victims showed different symptoms in different locations—a clue that led Osterhaus to suspect that the underlying cause must be a virus that suppresses the immune system. In time, the researchers found signs of infection with a distemper virus in the stricken seals—one similar but not identical to the lethal microbe that kills dogs and other members of the canine family.

At last it seemed that scientists had the answer to the appalling die-off, but some in the environmental community remained unconvinced. What had made the seals so vulnerable? Was it more than just coincidence that the disease had taken a much lighter toll along the less polluted shores of Scotland?

The early 1990s: The Mediterranean Sea

Although fishermen and yachtsmen who venture offshore sometimes encounter schools of striped dolphins surfing playfully in the bow wakes of boats, these small, sprightly, high-jumping cetaceans generally live and die out of human sight far from land. For this reason, the massive die-off in the Mediterranean population was well under way before wildlife researchers realized that yet another marine mammal had been hit by some deadly plague.

The first few dead and dying striped dolphins came ashore one by one near Valencia on Spain's eastern coast in July 1990, causing no suspicion that they were anything but isolated natural deaths. By mid-August, however, significant numbers of dead animals began hitting the beaches—not only around Valencia, but in Catalonia to the north and on Mallorca and the other Balearic Islands—as the disease ripped through the dolphin schools inhabiting the deep, open waters a dozen miles offshore. Physical examinations showed that the plague victims suffered from partially collapsed lungs, breathing difficulties, and abnormal movement and behavior. By the end of September, the death toll was rising along the French coast, and sick animals were also beginning to wash up in Italy and Morocco. But as winter set in, the epidemic slowed and finally stopped.

The following summer, the virulent disease broke out again in southern Italy and move eastward toward the western edge of the Greek islands. In the spring of 1993, it resurfaced in the Greek islands and spread to the east and northeast, claiming more and more victims as it went.

By the time the epidemic had played itself out, the official body count totaled more than eleven hundred. For every victim that washed ashore, several more vanished into the deep.

Once again, the killer proved to be a virus in the distemper family, but researchers found signs that contamination may have also played a role in the die-off.

Beginning in 1987, Alex Aguilar, a marine mammal specialist from the University of Barcelona, had been collecting fat samples from striped dolphins off northeastern Spain by firing specially designed darts from a crossbow or spear gun into animals riding the bow wakes.

When he compared these samples with those taken from the beached carcasses, he found that the victims of the die-off carried PCB levels two to three times higher than what he had found in healthy dolphins.

1992: Copenhagen, Denmark

Even a high school biology student could see the deformities in the tiny tadpolelike human sperm as they swam about under the microscope. In a single sample, some sperm might have two heads and others two tails, while another might have no head at all. Many didn't seem to swim right, showing total inactivity or frenetic hyperactivity instead of a strong, steady motion.

Over the years, Niels Skakkebaek, a reproductive researcher at the University of Copenhagen, had seen more and more sperm abnormalities, as well as a drop in the typical sperm count. At the same time, the rate of testicular cancer had tripled in Denmark between the 1940s and the 1980s. Skakkeback also noticed low sperm counts and unusual cells in the testes of men who eventually developed this type of cancer. Were the two findings connected?

Skakkebaek began to research the scientific literature, looking for other studies on sperm count, especially for data on men who did not suffer from infertility or other health problems. He and his colleagues eventually reviewed sixty-one studies, most of them from the United States and Europe, but also from India, Nigeria, Hong Kong, Thailand, Brazil, Libya, Peru, and Scandinavia.

The researchers were stunned by what they found. According to the data, average human male sperm counts had dropped by almost fifty percent between 1938 and 1990. At the same time, the incidence of testicular cancer had jumped sharply, not just in Denmark but in other countries as well. The medical data also suggested that genital abnormalities such as undescended testicles and shortened urinary tracts were on the rise among young boys.

Because the changes in sperm counts and quality and the increase of genital abnormalities had occurred over such a short period of time, the researchers ruled out genetic factors. Instead, the changes appeared due to some sort of environmental factor.

* * *

Beginning in the 1950s, these bizarre and puzzling problems began to surface in different parts of the world—in Florida, the Great Lakes, and California; in England, Denmark, the Mediterranean, and elsewhere. Many of the disturbing wildlife reports involved defective sexual organs and behavioral abnormalities, impaired fertility, the loss of young, or the sudden disappearance of entire animal populations. In time, the alarming reproductive problems first seen in wildlife touched humans, too.

Each incident was a clear sign that something was seriously wrong, but for years no one recognized that these disparate phenomena were all connected. While most incidents seemed linked somehow to chemical pollution, no one saw the common thread.

Then in the late '80s, one scientist began to put the pieces together.

2

HAND-ME-DOWN POISONS

FROM THE CORNER OF HER EYE, THEO COLBORN CAUGHT SIGHT OF another scientific paper shooting across the floor and settling on the carpet. She didn't even bother to turn her head. In hopes of slowing the tide of paper that was swamping her office in the fall of 1987, she had taken to closing her door, but it had done little good. With the wrist action of a Frisbee champion, the project director simply flicked them under the door. He had become so adept, he could wing a document to the center of the office. She let it sit there.

Flick. Another. Flick. Flick. Two more.

Sometimes, half a dozen documents would arrive in a single hour. She barely had time to file the reports and papers that bore titles such as "A Quantitative Assessment of Thyroid Histopathology of Herring Gulls (*Larus argentatus*) from the Great Lakes and a Hypothesis on the Causal Role of Environmental Contaminants," much less read them. Colborn looked around at the listing stacks of reports and studies and the bulging cardboard file boxes spread across the floor. So much had accumulated since she had begun her review of scientific papers concerning the health of wildlife and humans in the

Great Lakes region that the place was beginning to look like a landfill. If it got much worse, she would have to apply for a permit. Working late into the evening and on weekends had not solved the problem. As hard as she was scrambling, she felt as if she were getting buried.

How was she ever going to make sense of all this stuff?

Despite two months of full-tilt work, she still could not get a clear picture of how well the Great Lakes were recovering from decades of acute pollution. Seeking to learn all she could, she had gathered hundreds of papers, in addition to the ones that kept arriving across the threshold. One couldn't complain about a dearth of studies, but nothing coherent seemed to be emerging from her intensive research. What she had seemed like a hodgepodge of disconnected information, yet at the same time, she sensed that something important was lurking beneath the confusing surface. The most promising material seemed to be new data linking toxic chemicals to cancer in fish. That made sense. That is what one would expect to find in lakes reputed to be full of cancer-causing chemicals.

But how did the hundreds of other studies reporting all manner of strangeness fit into the picture? Why were the terns in polluted areas neglecting their nests? And what about the bizarre wasting syndrome observed in tern chicks, which seemed normal at first but then suddenly began losing weight until they withered away and died? Then there were the reports of female herring gulls nesting together instead of with males.

But even when the task seemed totally overwhelming, Colborn still felt lucky. Landing the job as scientist on this project to assess the environmental health of the Great Lakes had been a real break. She had joined the team at the Conservation Foundation, a nonprofit think tank in Washington, in early August 1987, feeling rather proud of herself and her new career. After all, she had come to Washington for the first time in her life just two years earlier, a fifty-eight-year-old grandmother with a brand-new Ph.D in zoology from the University of Wisconsin. Her first stint had been as a Congressional Fellow at the Office of Technology Assessment, a policy analysis group that did studies at the request of Congress (until abolished by the Republican majority in 1995), where she had worked on studies related to air pollution and water purification. Then the Conservation

Foundation had approached her about the Great Lakes study—an effort the group was undertaking with a Canadian counterpart, the Institute for Research on Public Policy.

This certainly beat filling prescriptions at a small-town drugstore in Carbondale, Colorado. Faced at fifty with the question of what she was going to do with the rest of her life, Colborn had briefly considered picking up her lapsed career as a pharmacist and opening a pharmacy in Carbondale, or she could have continued farming sheep, the career she had pursued since moving to Colorado from New Jersey fifteen years earlier. Either would have been the sensible course, but she decided instead to finally do what she had long desired.

Through her lifelong passion for watching birds, she had been drawn into the growing environmental movement and had spent years working as a volunteer on western water issues. Despite all she had learned on the front lines of various environmental battles, she still felt handicapped by a lack of official credentials. Without a degree behind you, it was easy for opponents to dismiss you as a do-gooder, a "little old lady in tennis shoes," even though she was tall, middle-aged, and shod in cowboy boots. Of course, that wasn't the whole story; her intellectual appetite had been whetted as well by what she had been able to learn on her own. So at fifty-one, Colborn plunged into a new life as a graduate student trudging up and down the mountains on the western slope of Colorado to take water samples for her master's work in ecology—a study of whether aquatic insects such as stone flies and mayflies could serve as indicators of river and stream health. Although some of her male advisers had been skeptical about investing energy in such an old graduate student, she had persisted and continued on to get her Ph.D.

Flick. Another document landed a foot from her chair—a copy of a speech by a governor from one of the states bordering on the Great Lakes. Like many of the other speeches and reports, this one touted the improvements in the Great Lakes and the signs of recovery. Public officials on both sides of the border had all but declared total victory in the battle against the severe pollution that had brought the Great Lakes nationwide notoriety in the 1960s and 1970s.

The nadir had been the day in June 1969, when the Cuyahoga River, which empties into Lake Erie in Cleveland, actually caught fire and burned a bridge. Six months later, when Cleveland's mayor recounted the fire before a congressional hearing on the proposed Clean Water Act, the city's burning river made national headlines. During this period, Lake Erie had been pronounced "dead" by the news media, and scientists judged the other lakes to be seriously endangered. At the height of this degradation, huge, stinking mats of rotting algae covered the beaches, bays and rivers were awash in oil and industrial waste, and once-abundant bird and wildlife populations collapsed.

The improvements since then were undeniable. Over two decades, the pea-soup algae and egregious pollution had gradually abated as local communities constructed sewage treatment plants, states imposed bans on the phosphates in detergents that had fueled runaway algal growth, and industries changed their practices to meet new limits on what they could discharge into water bodies. Following the federal restrictions on the pesticide DDT in 1972, the problem of thinning and broken eggshells that had devastated the bald eagles and other birds began to disappear, and many badly depleted bird populations rebounded dramatically. In fact, the numbers of some previously threatened birds, such as the herring gull and the double-crested cormorant, had climbed to an all-time high and now constituted a nuisance in many places. Yet, such explosive increases in opportunistic species could be something other than a sign of recovery for the lakes. Paradoxically, such a population explosion could signal a stressed ecosystem just as much as a population crash.

After two months of immersion in the wildlife literature and lengthy conversations with biologists working in the Great Lakes, she now had a strong gut feeling that the proclamations of recovery were premature. She had come to doubt that the lakes, however improved, were truly "cleaned up." The broken eggs littering the bird colonies may have disappeared, but biologists working in the field were still reporting things that were far from normal: vanished mink populations; unhatched eggs; deformities such as crossed bills, missing eyes, and clubbed feet in cormorants; and a puzzling indifference in usually vigilant nesting birds about their incubating eggs. Colborn's eyes wandered across the file boxes lined up the length of her office—forty-three of them, one for each species that had been stud-

ied in the Great Lakes. Everywhere in this wildlife data she was finding signs that something was still seriously wrong.

Whatever it was, the symptoms were not as visible and straightforward as those that had prompted Rachel Carson to write *Silent
Spring* almost a quarter of a century earlier. Carson, a scientist and
writer whose 1962 book helped spark the postwar environmental
movement, could graphically detail the damage done by wanton use
of man-made pesticides. It was hard to miss the masses of dead birds
that littered suburban backyards after aerial spraying in the 1950s or
to forget the image of a bird dying in a convulsive spasm from pesticide poisoning. Abnormal parental behavior or poor survival of young,
on the other hand, are less immediately apparent but perhaps no less
important in the long run for a species' survival.

And if something was still wrong, what were the implications
for humans living in the region? That was the ultimate question Colborn would be pursuing in the Great Lakes study.

She had already ordered public health reports from the United
States and Canada, and she had plans to survey data from several
cancer registries that had been set up in the Great Lakes basin in the
1970s—a time of increased public awareness of the pollution and
growing concern that the large numbers of toxic chemicals in the
lakes might be jeopardizing human health. There was a strong conviction among many people living in the region that they were being
exposed to higher levels of toxic chemicals than those living elsewhere and were suffering from higher than normal cancer rates.

The Great Lakes region appeared to be a good place to look for
links between environmental contaminants and cancer. Colborn
planned to scour the scientific literature and health statistics and
ferret out any clue. If there was anything there, she was determined
to find it. For the time being, she would set aside the puzzle of the
female gulls nesting together and the other distracting anomalies
and go after the cancer connection.

Her project director, Rich Liroff, was even more excited about
the cancer question than Colborn herself. As a member of the Science Advisory Board of the International Joint Commission, which
advises the United States and Canada on lakes policy, he had been

hearing a great deal about new findings on fish tumors. It was the hottest topic in Great Lakes research.

At his urging, Colborn was off a few weeks later to the Fourteenth Annual Aquatic Toxicity Workshop in Toronto, where the much anticipated final session would focus on chemical contaminants and fish tumors. Liroff's final instructions were to get some "ugly pictures" of cancer-ridden fish for the planned book that was to be based on their work.

The session on fish tumors more than lived up to expectations, as slides of grotesque fish tumors flashed across the screen in the darkened meeting room. Clearly this was key evidence that would give direction to her research on human and wildlife health, for the researchers were making a persuasive cause-and-effect link between specific chemicals found in the Great Lakes basin and cancer. Although common sense would suggest that the health problems in wildlife and the contamination that had become pervasive must be connected, it often proved impossible to pin an abnormality to a single chemical in part because the animals were being exposed to *hundreds* of chemicals, most of them still unidentified.

Some skeptics had questioned whether the cancer in fish was related to chemical contamination at all. But presentations in the fish tumor session put several alternative explanations to rest, including the suggestion that the cancer outbreaks were not a new phenomenon but a natural event caused by viruses.

John Harshbarger, a leading expert on cancer in wildlife from the Smithsonian Institution, had responded that historical data showed that large outbreaks of cancer in fish had been reported only since the chemical revolution of the past half century, which had poured countless tons of man-made chemicals into the environment. In all the outbreaks except one, he added, viruses had been excluded as a cause.

The documented outbreaks had also followed a distinct pattern. The fish cancers showed up below discharge pipes from industrial operations or municipalities and affected fish species that spend most of their time on the bottom mingling with the muds and sediment. Moreover, scientists had gone on to demonstrate the link between the contaminated sediments and the cancer in laboratory experiments using confined fish. When researchers fed contaminants extracted from the sediments to the fish or applied them to the skin, the fish developed the same cancers seen in the wild.

A series of other talks followed, which made the case that the chemicals causing the cancer were polyaromatic hydrocarbons, or PAHs, a class of chemicals found in petroleum products or created by the incomplete burning of any carbon-containing material ranging from gasoline to hamburgers on the outdoor grill.

A team from the U.S. Fish and Wildlife Service had ruled out not only viruses but also metals and synthetic chlorine–containing chemicals such as DDT in their studies of cancer in brown bullheads, a type of catfish, in the Black and Cuyahoga Rivers, which flow into Lake Erie. The slides of tumor-ridden fish from the notorious Cuyahoga River were a sorry sight, with deformed, knobby whiskers and oozing cancers on their smooth scaleless skin. The researchers had found high concentrations of PAHs in the tissue of these cancer-ridden fish and PAH breakdown products in the bile in their livers.

Other researchers reported on an international collaborative effort that was shedding light on how PAHs did their damage inside the body. Diseased fish showed changes in the liver caused by bonding between carbon-based organic chemicals such as PAHs and the DNA in the nuclei of their cells. This phenomenon of bonding between foreign chemicals and DNA, which carries an individual's genetic blueprint, had been associated with the earliest stage of cancers caused by chemicals.

Taken together, this was the most sophisticated work to date and something of a breakthrough. The studies—inducing the same cancer in the laboratory as seen in the wild, isolating PAHs in fish tissue, and showing chemically induced cancer-related changes at the cellular level—all showed that the link between fish cancer and contaminated sediments was more than a coincidental association.

Amid the buzz of excitement about the growing evidence linking PAHs to fish tumors, the meeting's keynote address added some sobering perspective.

Those trying to discover links between ill health and contaminants were losing the battle, declared Bengt-Erik Bengtsson, the head of the Swedish Environmental Protection Board's Laboratory for Aquatic Toxicology. Despite noteworthy advances, toxicologists were falling further and further behind in their ability to analyze and identify the contamination they encountered in the environment.

In the Baltic Sea, he noted, biologists had reported a reduction in the size of the testes of fish—a condition apparently related to the amount of contamination from organochlorines, a class of man-made chemicals containing chlorine. But they had been at a loss to discover which chemical was causing the problem, for current analytical methods had been able to identify only *six percent* of the synthetic organochlorine chemicals found in Baltic waters. Chemical companies were putting hundreds of new synthetic chemicals onto the market each year—far faster than toxicologists and regulatory agencies were able to develop new assays to detect them, Bengtsson warned. It was a wonder researchers had succeeded in linking anything—so many chemicals, so many effects.

Colborn intuitively sensed that Bengtsson's speech contained an important clue about how many chemicals were acting on wildlife in the Great Lakes and elsewhere. But because of her immediate focus on the cancer connection, she put it out of her mind. Months would pass before she came to recognize its full importance.

When Colborn returned from Toronto with ugly pictures in hand and renewed enthusiasm for the daunting task that lay ahead, the public health reports that she had ordered were waiting on her desk.

She dove into the health data, focusing especially on those areas where wildlife researchers had found cancer in fish. If the wildlife were providing warnings, she reasoned, then that was where one would first expect to find humans with higher cancer rates.

To her dismay, the cancer registries that had been set up in the Great Lakes basin proved useless, for none had been in operation long enough to yield conclusions about trends or comparative risk for those regions bordering the lakes. So Colborn turned to broader reports on human cancer in the United States and Canada. She pored over the computer printouts and reports for hours, analyzing the data from various perspectives to see if she could tease out meaningful patterns. When nothing emerged, she went at it again from a new direction. She looked for clusters of a single kind of cancer, higher overall cancer rates, striking geographical patterns in the cancer incidence, or anything else out of the ordinary.

Finally, after months of concerted effort, she had to admit that no matter which way she cut the data, they yielded no support for the belief that people in the Great Lakes basin were dying of cancer more than people elsewhere in the United States and Canada. Surprisingly, the opposite appeared to be the case. The rates for some cancers were actually *lower* in the Great Lakes area than in some other regions. There was simply no evidence in the public health records of elevated or unusual cancer patterns among those living near the lakes.

Colborn was puzzled. The high cancer rates she had heard so much about appeared to be more myth than reality. After months of chasing the specter of cancer, she found herself at a dead end.

Faced with this major setback, she turned her mind again to the wildlife literature and tried to think clearly about where she should go next. Sitting there surrounded by the boxes of animal studies, she was suddenly struck by the obvious. Why on earth hadn't she seen this before? The fish cancer research might be cutting-edge work, but most of the problems being reported in wildlife were *not* cancer. Except in fish in highly contaminated areas, cancer was an extreme rarity. Yet a long list of fish and animals all across the Great Lakes basin showed ill health that seemed to be impairing their survival.

The phrase is automatic: "cancer-causing chemical." The habit of mind is so ingrained, we do not even recognize the conceptual equation that has dominated our thinking about chemicals. For the past three decades, the words "toxic chemical" have become almost synonymous with cancer not only in the public mind but in the minds of scientists and regulators as well. Colborn was no exception. But now she recognized that this preoccupation with cancer and mutations had been blinding her to the diversity of data she had collected. Moving beyond cancer proved to be the most important step in her journey, for as she looked at the same material with new eyes, she gradually began to recognize important clues and follow where they led.

Colborn wasn't sure where to turn next. If the problem wasn't cancer, what was it? She was still floundering around in a morass of undigested information. By this time, she had collected several hundred scientific papers and dozens of studies and reports, but each new document seemed only to add to the confusion.

For want of a better idea, she decided to go back and again read through the files on mink, otter, fish, and birds such as the bald eagle and herring gull.

At the recommendation of her project director, Colborn had traveled earlier to Hull, outside Ottawa, to meet with Mike Gilbertson, Glen Fox, and other veteran researchers in the Canadian Wildlife Service, who had been investigating the problems in Great Lakes wildlife for more than a decade. The trip had been invaluable both for the information she had gathered and for the professional friendships that had developed from that initial encounter.

Gilbertson had given Colborn complete access to his meticulously organized collection of material on each animal species that breeds in the Great Lakes basin—data that he had gathered over the years and arranged in chronological order in three-ring binders. Colborn was awed by the elegance of the effort and by the years of dedication and scholarly consideration that it reflected. With a sense of history, Gilbertson had gone to great lengths to collect papers and studies dating back a half century or more—literature documenting that the problems seen today in the birds and wildlife along the lakes had not been reported before World War II. In the bald eagle file, she found evidence of parallel declines in the postwar period in the bald eagle in North America and in its European cousin, the white-tailed sea eagle, along with a collection of reports detailing the concentrations of synthetic chemical contaminants found in both species. Photocopies from Gilbertson's archive had greatly enriched Colborn's files, but their conversations, during which Gilbertson generously shared his broad experience, had proved even more invaluable.

Over lunch in the Canadian Wildlife Service cafeteria, Colborn, Gilbertson, and Fox had discussed the wildlife evidence contradicting the frequent claims that the lakes had been cleaned up. The two Canadians shared the conviction the wildlife work had likely implications for human health and constituted a warning humans ought to heed. In her survey of the scientific literature, Colborn had been fascinated by some of Fox's work, which reported evidence of behavioral changes in the wildlife as well as signs of physical damage.

In herring gull colonies, particularly in highly polluted areas of Lakes Ontario and Michigan, Fox and his colleagues had found nests with twice the normal number of eggs—a sign that the birds occupying the nests were two females instead of the expected male-female pair. The phenomenon, which persisted in some areas, had been particularly prevalent in the mid to late 1970s. During this period, Fox had collected and preserved seventeen near-term embryos and newly hatched chicks from the affected colonies in hopes that he might eventually discover what was causing this unusual behavior and other reproductive problems.

A few years later, Fox encountered a scientist who might help him find the answer. Michael Fry, a wildlife toxicologist at the University of California at Davis, had investigated how the pesticide DDT and other synthetic chemicals disrupt the sexual development of birds after hearing reports of nests with female pairs in western gull colonies in southern California. While some looked for an evolutionary explanation for the phenomenon, Fry had suspected contamination. Reports in the scientific literature indicated that a number of synthetic chemicals, including the pesticide DDT, could somehow act like the female hormone estrogen.

To test his theory, Fry had injected eggs taken from western gull and California gull colonies in relatively uncontaminated areas with four substances—two forms of DDT; DDE, the breakdown product of DDT; and methoxychlor, another synthetic pesticide that has also been reported to act like the hormone estrogen. The experiment showed that the levels of DDT reported in contaminated areas would disrupt the sexual development of male birds. Fry noted a feminization of the males' reproductive tracts, evident by the presence of typically female cell types in the testicles or, in cases of higher doses, by the presence of an oviduct, the egg-laying canal normally found in females. Despite all this internal disruption, the chicks had no visible defects and looked completely normal.

As soon as he could make arrangements, Fox shipped the preserved embryos and chicks off to Fry in California. In his examination of the birds' reproductive tracts, Fry found that five of the seven males were significantly feminized and two had visibly abnormal sex organs. Five of the nine females showed significant signs of disrupted development as well, including the presence of two egg-laying canals instead of the one that is normal in gulls. Such disruptions, Fry noted, could indicate that the birds had been exposed to chemicals that acted like the female hormone estrogen.

Earlier experiments by other researchers had shown that exposing male birds to estrogen during development affects the brain as well as the reproductive tract and permanently suppresses sexual behavior. When chicken and Japanese quail eggs received estrogen injections, the males that hatched never crowed, strutted, or exhibited mating behavior as adults.

Taken together, the evidence in the Great Lakes suggested that the females were nesting together because of a shortage of males, which might be absent because they were disinterested in mating or incapable of reproducing. Though most of the eggs in these same-sex nests were infertile, these females sometimes managed to mate with an already paired male and hatch a chick. The female pairs appeared to be an effort to make the best of a bad situation.

Fox and others had noticed other behavioral abnormalities as well, particularly in birds that had high levels of chemical contamination. In Lake Ontario colonies, the birds showed aberrant parental behavior, including less inclination to defend their nests or sit on their eggs. In unsuccessful nests, the incubating eggs were unattended for three times as long as in nests where birds successfully produced offspring. A study comparing reproduction in Forster's terns nesting in clean and contaminated areas reported that nest abandonment and egg disappearance, often due to theft by predators, was substantial in the contaminated area on Lake Michigan but virtually nonexistent in the clean colony on a smaller lake in Wisconsin. Parental inattentiveness clearly diminished the chances that the eggs would hatch and the chicks would survive.

What Colborn remembered afterward about the conversation was how cautious they had all been. Despite the shared view that wildlife findings had implications for humans, no one wanted to acknowledge the unspoken question hanging in the air. No one dared ask whether synthetic chemicals might be having similar disrupting effects on human behavior. Those were treacherous waters they all preferred to avoid.

As Colborn tackled the wildlife files for a second time, her mind kept returning to the female gulls nesting together. She pulled out the papers by Fox and Fry and carefully reread them. She sensed that the "gay gulls," as someone had dubbed them, were an important piece of the puzzle, but she still didn't know how to put it all together. The feminization of the males was a consequence of disrupted hormones. That involved the endocrine system, which was composed of various glands that controlled critical functions such as basic metabolism and reproduction.

Well, that about summed up her knowledge of current endocrinology. She had taken courses in pharmacy school, but the intervening decades had revolutionized the field. And endocrinology was not standard fare in the training of ecologists. If she was going to pursue this line of inquiry, she would have to know more.

Several new endocrinology textbooks joined the stacks of wildlife files on the top of her desk. Her first efforts to master the basics of the endocrine system proved frustrating in the extreme. The texts were dense, unreadable, and full of acronyms that forced one to keep flipping back to earlier pages. Colborn only began making headway when she found a practical, accessible text, *Clinical Endocrine Physiology*, which she kept within reach through the months that followed.

As she focused on hormones, evidence that she had previously passed over gained new meaning. She recalled the keynote address by Bengtsson, the Swedish toxicologist who described how the size of fish testicles had diminished as contamination from synthetic organochlorine chemicals increased in the Baltic. Was this a sign of hormone disruption? She looked again at reports of abnormal mating behavior in bald eagles, which had preceded the appearance of eggshell thinning and the collapse of the eagle population. The birds had been disinterested in mating. Hormone disruption, Colborn now suspected.

Other things struck her, too, as she read through the wildlife files. A pattern began to emerge. Birds, mammals, and fish seemed to be experiencing similar reproductive problems. Although the adults living in and around the lakes were reproducing, their offspring often did not survive. Colborn began to focus on studies that compared Great Lakes populations to others living inland. In every case, the lake dwellers, who appeared otherwise healthy, were far less success- ful in producing surviving offspring. It seemed that the contamina- tion in the parents was somehow affecting their young.

It dawned on Colborn that the human studies investigating the effects of exposure to synthetic chemicals had focused largely on can- cer in exposed adults. Only a handful had looked for possible effects on the children of exposed individuals, but Colborn recalled reading one that had studied the children of women who had regularly eaten Great Lakes fish. She dug it out of her files and read it again. The study by Sandra and Joseph Jacobson, psychologists from Wayne State University in Detroit, had also found evidence the mother's level of chemical contamination affected her baby's development. The children of mothers who had eaten two to three meals a month of fish were born sooner, weighed less, and had smaller heads than those whose mother did not eat the fish. Moreover, the greater the amount of PCBs, a persistent industrial chemical that is a common pollutant in Great Lakes fish, in the umbilical cord blood, the more poorly the child scored on tests assessing neurological development, lagging behind in various measures, such as short-term memory, that tend to predict later IQ.

The parallel between this human study and the offspring effects in wildlife was interesting as well as troubling.

Colborn moved on, following wherever the investigation led, but the Jacobson study hovered in the back of her mind, nagging at her like an unanswered question. Had scientists been looking in the right place in their search for effects? Perhaps the Jacobson studies were more important than anyone had realized.

As she dug deeper, more parallels became apparent. In the tissue analyses done on the wildlife, the same chemicals kept showing up in the troubled species, among them the pesticides DDT, dieldrin, chlor- dane, and lindane, as well as the family of industrial chemicals called PCBs, which had been used in electrical equipment and many other products. Of course, these results might be a coincidence, or they could well reflect technical limits and the small budgets for tracking contaminants. These were the chemicals toxicologists knew how to measure and the ones least expensive to analyze.

Whatever the reason for their repeated appearance, studies had found the very same chemicals in human blood and body fat. Colborn was particularly shocked by the high concentrations reported in the fat in human breast milk.

By the time the research deadline approached, Colborn had plowed through more than two thousand scientific papers and five hundred government documents. She felt like a beagle following its nose. She wasn't sure where she was headed, but propelled by her curiosity and intuition, she was hot on the trail. She had found so many tantalizing parallels, so many echoes among the studies. Somehow, she was certain, it all fit together, because she kept finding unexpected links. Her latest discovery had come while she reexplored the literature on the eerie wasting syndrome seen in young birds. The chicks could look normal and healthy for days, but then suddenly and unpredictably they would begin to languish, waste away, and finally die. The wasting problem, scientists were learning, was a symptom of disordered metabolism. The young birds could not produce sufficient energy to survive. Though one would not at first suspect that this problem had anything in common with the gay gull phenomenon, it also stemmed from the disruption of the endocrine system and hormones.

But the elation of discovery passed quickly. The deadline was looming. What did this all mean? She had pieces and patterns but no picture.

Maybe she could gain perspective if she laid it all out. Colborn began entering the findings from the studies on a huge ledger sheet, the kind used by accountants. When that became unwieldy, she turned to her computer and created an electronic spreadsheet, which scientists call a matrix. As she made entries under columns headed "population decline," "reproductive effects," "tumors," "wasting," "immune suppression," and "behavioral changes," her attention came to focus increasingly on sixteen of the forty-three Great Lakes species that seemed to be having the greatest array of problems.

She sat back and looked at the list: the bald eagle, lake trout,

herring gull, mink, otter, and the double-crested cormorant, the
snapping turtle, the common tern, and coho salmon. What did they
have in common?

Of course! Each and every one of these animals was a top
predator that fed on Great Lakes fish. Although the concentrations
of contaminants such as PCBs are so low in the water in the Great
Lakes that they cannot be measured using standard water testing
procedures, such persistent chemicals concentrate in the tissue and
accumulate exponentially as they move from animal to animal up
the food chain. Through this process of magnification, the concentrations of a persistent chemical that resists breakdown and accumulates in body fat can be 25 million times greater in a top predator
such as a herring gull than in the surrounding water.

One other startling fact emerged from the spreadsheet. According to the scientific literature, the *adult* animals appeared to be doing
fine. The health problems were found primarily in their offspring. Although she had been thinking about offspring effects, Colborn had
not recognized this stark, across-the-board contrast between adults
and young.

Now the pieces were beginning to fall together. If the chemicals
found in the parents' bodies were to blame, they were acting as
hand-me-down poisons, passed down from one generation to the
next, that victimized the unborn and the very young. The conclusion
was chilling.

But the host of disparate symptoms in everything from adult
herring gulls to baby snapping turtles did not seem to add up. Some
animals, like the gulls, exhibited strange behavior such as same-sex
nests, while other species, including the double-crested cormorant,
had visible gross birth defects such as club feet, missing eyes,
crooked spines, and crossed bills. Again a pattern emerged from the
confusing pieces of the puzzle as Colborn reflected on what she had
learned by following her nose.

These were all cases of derailed development, a process guided to
a significant extent by hormones. Most could be linked to disruption
of the endocrine system.

This insight pointed Colborn's investigation in another direction. She began reading everything she could find about the chemicals

Lake Ontario Biomagnification of PCBs

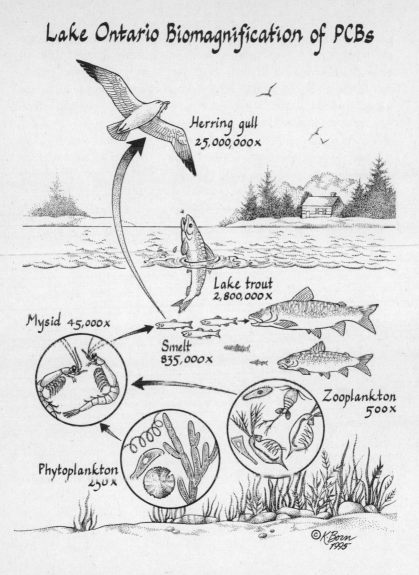

Herring gull
25,000,000x

Lake trout
2,800,000x

Mysid 45,000x

Smelt
835,000x

Zooplankton
500x

Phytoplankton
250x

©K.Born
1995

As PCBs work their way up the food chain, their concentrations in animal tissue can be magnified up to 25 million times. Microscopic organisms pick up persistent chemicals from sediments, a continuing source of contamination, and water and are consumed in large numbers by filter feeding tiny animals called zooplankton. Larger species like mysids then consume zooplankton, fish eat the mysids, and so on up the food web to the herring gull.

that showed up again and again in the tissue analyses of animals having trouble producing viable young. She quickly learned that the testing and reviews done by manufacturers and government regulatory agencies had focused largely on whether a chemical might cause cancer, but she found enough in the peer-reviewed scientific literature to prove that her hunch had been correct.

The hand-me-down poisons found in the fat of the wildlife had one thing in common: one way or another, they all acted on the endocrine system, which regulates the body's vital internal processes and guides critical phases of prenatal development. The hand-me-down poisons disrupted hormones.

3

CHEMICAL MESSENGERS

PUSHING ON WITH HER RESEARCH ON HORMONES, THEO COLBORN DIS-
covered a central piece of the puzzle in the world of Frederick vom
Saal, a biologist at the University of Missouri. Vom Saal's exploration
of how hormones help make us who we are is a fascinating scientific
adventure in its own right. In a series of experiments with mice, he
showed that small shifts in hormones before birth can matter a great
deal and have consequences that last a lifetime. His work helped
highlight the hazard posed by synthetic chemicals that can disrupt
hormonal systems.

Vom Saal's investigation of the wondrous world of hormones
began in 1976 during his postdoctoral days at the University of Texas
in Austin, inspired by the behavior of the lab mice. Like most postdoctoral biology students, vom Saal was spending the better part of
his life in the lab, where his regular chores included breeding mice.
As he played mouse matchmaker, arranging encounters between
eager males and receptive females, he became intrigued by the inter-
play between the animals as he moved them from cage to cage.

In the beginning, the small, white, pink-eyed creatures had all seemed like cookie-cutter copies of each other. But as he watched the females scurrying about in the breeding cages, individuals quickly emerged from the crowd. Whenever he returned a female to a group cage holding half a dozen females, there always seemed to be one mouse who would attack the intruder. These were mice with an attitude—tough cookies who rattled their tails threateningly and lashed out at their mild-mannered companions.

Such a difference between the behavior of one female and another was striking—and puzzling. The mice were all from a single laboratory strain that had been inbred for generations. When it came to genes, they were virtually identical.

This simple observation set the course for vom Saal's life's work in reproductive biology. In the years that followed, he designed dozens of experiments to probe the mystery of how two mice with almost the same genetic blueprint could behave so differently.

The notion persists that genes are tantamount to destiny and that one might explain everything from cancer to homosexuality by locating the responsible genes. But in a series of scientific papers, vom Saal demonstrated that there are other powerful forces shaping individuals—females as well as males—before birth. Genes, it turned out, are not the whole story. Not by a long shot.

What vom Saal saw during those long hours observing mice in the lab contradicted *everything* he had read. According to the scientific literature of the period (which reflected prevailing human assumptions as much as it described animal behavior), aggression was strictly a male behavior. But if tail-rattling, chasing, and biting among the females weren't aggression, what *would* one call it?

Eventually, vom Saal's colleagues had to concede that the behavior did look like aggression, but they tended to shrug it off as unimportant. Males were the center of the action in animal societies according to the prevailing wisdom in the field of animal behavior, so what females did simply didn't matter. They were just passive baby makers.

Vom Saal wasn't so sure. His intuition told him what he was seeing was probably important as well as interesting. His doctoral work had centered on the role played by testosterone in development before birth, and he knew that this hormone—found at much higher levels in males—drives aggression.

From his observations, the tough females weren't common, but they weren't rare either. There seemed to be roughly one aggressive female for every six mice in the colony—something he noticed because the mice were housed six to a cage. If the mice were clones, something besides genes had to be shaping the aggressive females. Since birth the sisters had been raised identically, so living conditions could not explain the differences. Could the cause be something in their prenatal environment?

That set him to thinking about how mice are carried before birth. Their mother's womb isn't a single compartment like the human womb, but two separate compartments or "horns" that branch off to the left and the right at the top of the vagina or birth canal. The baby mice are tucked in the narrow horns like peas in a pod—as many as six on a side. This arrangement means that some of the females will develop sandwiched between two males.

Vom Saal began calculating probabilities. If there were twelve mice in the typical mouse litter and if the placement of males and females in the womb was random, how many females would end up between two males? Roughly one in six, he figured. That supported the theory taking shape in his head. Some of the females are markedly more aggressive, he suspected, because they had spent their prenatal life wedged between two males. A week before birth, the testicles in a male pup begin to secrete the male hormone testosterone, which drives his own sexual development. The female pups might be bathed in testosterone washing over from their male neighbors.

Maybe, vom Saal thought, the answer to the mystery of how genetically identical females could be so different lay in hormones—chemical messengers that travel in the bloodstream, carrying messages from one part of the body to another.

In the body's constant conversation with itself, nerves are just one avenue of communication—the one employed for quick, discrete messages that direct a hand to move away from a hot stove. A large part of the body's internal conversation, however, is carried on through the bloodstream, where hormones and other chemical messengers move about on the biological equivalent of the information superhighway, carrying signals that not only govern sex and reproduction but also coordinate organs and tissues that work in concert to keep the body functioning properly.

Hormones, which get their name from the Greek word meaning "to urge on," are produced and released into the bloodstream by a variety of organs known as endocrine glands, including the testicles, the ovaries, the pancreas, the adrenal glands, the thyroid, the parathyroid, and the thymus. The thyroid, for example, produces chemical messengers that activate the body's overall metabolism, stimulating tissues to produce more heat. In addition to eggs, a woman's ovaries release estrogens—the female hormones that travel in the bloodstream to the uterus, where they trigger growth of the tissue lining the womb in anticipation of a possible pregnancy.

Yet another endocrine gland, the pituitary, which dangles on a stalk from the underside of the brain just behind the nose, acts as a control center, telling the ovaries or the thyroid when to send their chemical messages and how much to send. The pituitary gets its cues from a nearby portion of the brain called the hypothalamus, a teaspoon-size center on the bottom of the brain that constantly monitors the hormone levels in the blood in much the way that a thermostat monitors the air temperature in a house. If levels of a hormone get too high or too low, the hypothalamus sends a message to the pituitary, which signals the gland that produces this hormone to gear up, slow down, or shut off.

The messages travel back and forth continuously. Without this cross talk and constant feedback, the human body would be an unruly mob of some 50 trillion cells rather than an integrated organism operating from a single script.

As scientists have delved deeper into the nervous, immune, and endocrine systems—the body's three great integrating networks— they have encountered profound interconnections: between the brain and the immune system, the immune system and the endocrine system, and the endocrine system and the brain. The links sometimes seem utterly mystifying. How, for example, could a woman suffering from multiple personality disorder play with a cat for hours while she was one personality and suffer violent allergic reactions to cats when she took on another?

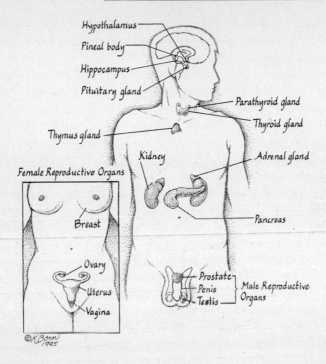

Hypothalamus
Pineal body
Hippocampus
Pituitary gland
Parathyroid gland
Thyroid gland
Thymus gland
Kidney
Adrenal gland
Female Reproductive Organs
Breast
Pancreas
Ovary
Uterus
Vagina
Prostate
Penis
Testis
Male Reproductive Organs

©KBorn
1995

Some important glands, organs, and tissues sending or receiving hormonal messages in the human body.

Nobody knows the answer to this question, but it certainly lies in this internal conversation and the constant babble of chemical messengers. Changes in one part of this complex, interconnected system can have dramatic and unexpected consequences elsewhere, often where one might least expect, because everything is linked to everything else. A brain tumor, for example, might show up as disrupted menstrual cycles and hypersensitivity of the skin rather than as headaches.

If hormones are vital to maintain proper functioning in adults, they are perhaps even more important in the elaborate process of development before birth.

But how could vom Saal test his theory?

Mouse Caesarean sections.

Just before the females were ready to give birth at the end of their nineteen-day pregnancies, vom Saal removed the tiny babies, who were approximately an inch long and about the size of an olive. He marked them based on their position relative to their neighbors in the womb. In this way, he could discover where aggressive females had spent their prenatal lives. Thus began vom Saal's exploration of what some in the field playfully refer to as the "wombmate" effect, known formally as intrauterine position phenomenon.

Although vom Saal is now forty-nine and a professor at the University of Missouri, he still looks youthful enough to be mistaken for a graduate student. In a scientific world where many seldom venture beyond narrow specialities, vom Saal embraces the big picture, unabashedly declaring that he is interested in "womb-to-tomb biology." He moves easily between elegant, tightly focused studies and a larger, more encompassing pursuit of fundamental questions: Why does this happen? What is the evolutionary significance?

Those first studies in Austin confirmed his theory. As the mice removed by Caesarean section matured, the aggressive females were, as predicted, the ones who had developed between brothers. Each intriguing finding raised new questions, leading to more studies and, in time, observations on thousands of mice delivered by Caesarean section. Aggression proved just the most obvious sign of profound differences between mouse sisters that could be predicted to a remarkable degree by their position in the womb.

At first blush, vom Saal's results sound like a tale of the ugly sister and the pretty sister. Not only was the ugly sister—the mouse that had developed between males—more aggressive, but vom Saal discovered she was significantly *less* attractive to males than the pretty sisters who had spent their womb time between other females. Eight times out of ten, a male given a choice would chose to mate with the pretty sister.

What's attractive to males isn't the female's tiny pink eyes or the curve of her tail. The social life of mice is governed by the nose, and the attractiveness of females depends on the social chemicals they give off, which are called pheromones. The pretty sisters smell "sexier" to males because they produce different chemicals than their less attractive sisters. The prenatal hormone environment leaves a

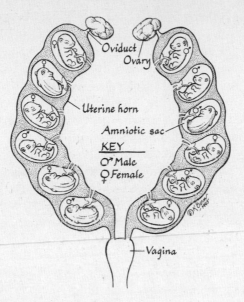

Behavioral and reproductive differences in mice can be predicted to a re-markable degree by their position, which is related to hormone exposure, in the womb. (Adapted from vom Saal and Dhar, 1992)

permanent imprint on each sister that is recognized by males for the rest of her life.

The sisters also showed dramatic differences in their reproduc-tive cycles. Besides finding mates more readily, the pretty sister also matured faster than her ugly sister and came into heat—a period of sexual receptivity—more often. As a consequence, she had more op-portunities to get pregnant and was more likely overall to produce more offspring in her lifetime than her aggressive, unattractive sister, who experienced puberty later and came into heat less frequently.

Even more amazing, studies by other researchers, including Mertice Clark, Peter Karpiuk and Bennett Galef of McMaster Uni-versity, and the team of John Vandenbergh and Cynthia Huggett of North Carolina State University, have found that the wombmate ef-fect even influences whether a female will give birth to more males or more females when she has pups of her own. This is mysterious in-deed, since scientists up to now believed that the mother has no role

in determining the sex of her offspring. Based on current understanding, it is the sperm contributed by the father that dictates
whether the egg develops into a male or female, so how a mother influences sex ratio is still unknown. However it happens, the pretty
sisters tend to have litters made up of sixty percent females, while
the ugly sisters generally give birth to litters that are roughly sixty
percent male. As Vandenbergh wrote of this transgenerational
wombmate influence: "Brothers beget nephews."

After hearing the tale of the two sisters, one might easily conclude that it would be wise to be a pretty sister if one had to be a
mouse. They have lots of mates and babies and, judged by the evolutionary imperative of producing offspring, seem more successful
than their ugly sisters.

Not so fast, vom Saal cautions. When one considers how these
sisters live their lives within a mouse population that goes through
boom and bust cycles, the pretty sister begins to lose her obvious
edge. Typically, a mouse population builds to a very high peak and
then it crashes. In ordinary times when the population isn't too
dense, the pretty sisters definitely have the advantage, but as conditions become overcrowded the pretty sisters' ability to produce babies diminishes because the females respond to scent cues in urine
that inhibit reproduction.

But these overcrowded times are precisely when the ugly sisters
come into their own. Because they are relatively immune to the inhibiting cues, they are likely to be the only ones to produce offspring,
and the ugly sisters are the only ones tough enough to protect their
babies from attack and infanticide.

Interestingly, some studies have shown that the mother's physical condition can also alter hormone levels in the womb and influence the offspring. Mouse mothers that experience continuous
stress through the latter part of their pregnancies give birth to females who have all the physical and behavioral characteristics of
females who develop between males. Maternal stress seems to override the ordinary wombmate variations and produce a litter composed solely of tough cookies.

So what's the evolutionary lesson in this tale?

In vom Saal's view, the real lesson is the value of variability.

The acute sensitivity of developing mammals such as mice to slight shifts in hormone levels in the womb has been shaped by evolution. This characteristic helped insure wide variation in the offspring, even wider variation than that produced by genetic shuffling alone. Variation is the way mammals have hedged their bets in the face of a rapidly shifting environment. If you don't know what the conditions will be for your offspring, the best thing to do is produce many different kinds in the hope that at least one of them will be suited to the emerging moment.

Vom Saal's early investigations into the wombmate effect focused solely on females. The decision to look at males to see if female wombmates had any influence on them was almost an afterthought. Though the results would round out this line of research, vom Saal admits he frankly did not expect to find anything remarkable. It was widely assumed that male development was driven exclusively by testosterone, so being next to females should make little difference.

In fact, the results of his experiments astonished him. The wombmate effect shaped the destinies of males as well as females and in ways that no one would have *ever* predicted. In a major paper in the prestigious journal *Science* in June 1980, vom Saal and his associates laid out the case that it was exposure to the female hormone estrogen before birth that *increased* a male's sexual activity in adult life.

Inside and outside the world of science, many have regarded the level of male sexual activity as an index of masculinity and a product of the male hormone testosterone. Indeed, the findings were so counterintuitive and so contrary to assumptions about the "male" hormone testosterone and the "female" hormone estrogen that one of his collaborators protested that they must have somehow mixed up the samples. Vom Saal found, however, that estrogen and testosterone each influence males—and in ways that run counter to our conventional notions of "maleness" and "femaleness." The effect of wombmates on males proved an even more provocative vein of research than his earlier work on females.

If the females seem a story of the pretty and ugly sisters, then vom Saal's findings on the males sound like a tale of the playboy and the good father.

As adults, the playboy males, exposed to higher levels of estrogen by their female wombmates, showed another surprising characteristic besides their higher rates of sexual activity. It would seem logical to assume that exposure to estrogen might make males more solicitous toward the young, but in fact, the opposite proved true. When placed with young mice, these males were more likely to attack and kill babies. The high-testosterone males who had had brothers for wombmates turned out to be the good daddies, who surprisingly showed almost as great an inclination to take care of pups as mouse mothers.

The playboy males were standouts in one other respect as well—the size of their prostate, the small gland that wraps around the urethra, through which urine is eliminated. The males exposed to higher levels of estrogen had prostates that were fifty percent larger than those seen in brothers who had had male wombmates. In addition, these larger prostates are more sensitive to male hormones in adulthood because they contain three times the number of testosterone receptors found in the prostates of brothers with male wombmates. More receptors generally means that the gland will grow more quickly in response to male hormones circulating in the bloodstream in adulthood.

Although human babies don't usually have to share the womb with siblings, their development can nevertheless be affected by varying hormone levels, which occur in the womb for reasons scientists don't completely understand. Medical problems such as high blood pressure can drive up estrogen levels, for example. Or perhaps eating tofu, alfalfa sprouts, or other foods that are high in plant estrogens during pregnancy could boost estrogen exposure. There is also the possibility that the mother's body fat contains synthetic chemicals that disrupt hormones.

Whatever the source, a recent study on opposite-sex human twins showed that wombmate effects can be detected in people as well. The study, which focused on an obscure difference in the auditory systems of males and females that exists from birth, found that girls who had developed with a boy twin showed a male pattern, suggesting that they, like vom Saal's female mice, had been somewhat masculinized by the hormones spilling over from a male wombmate.

In the midst of all these surprises, the male wombmate studies in mice yielded only one expected result—on male aggression. Males with male wombmates and the highest testosterone exposure before birth were indeed the most aggressive toward other adult males, and males with female wombmates were the least aggressive.

Scientists working in this field are still debating how estrogen shapes the development of males and females, particularly the development of the brain and behavior, but vom Saal believes that estrogen is helping to masculinize males by acting to enhance some effects of the male hormone testosterone. Together the two hormones influence the organization of the developing brain to increase the level of sexual activity the male mouse will exhibit as an adult. Vom Saal had demonstrated that this is a prenatal effect rather than a consequence of adult hormone levels by castrating the mice shortly after birth and then in adulthood administering an identical amount of male hormone to brothers with male and female wombmates. Even with identical hormone exposure these male mice showed different levels of sexual activity—evidence that adult hormone levels are not the cause of these behavioral differences.

Those who hear about vom Saal's work typically ask him, Which is the "normal" mouse: the pretty sister or the ugly sister? The playboy or the good father?

"They're *all* normal," vom Saal says emphatically.

The question itself seems to stem from our dualistic notion of maleness and femaleness, which sees the two sexes as mutually exclusive categories. In fact, there are many shades of gray and overlap between behaviors thought of as typically male or female. Seen in this light, there is nothing abnormal about an aggressive female or a nurturing male. In this strain of mice, whose genetic variability has been reduced by generations of inbreeding, these individuals reflect the variability created by the natural influence of hormones before birth. What is "normal," vom Saal says, returning to an evolutionary theme, is not one type of individual or another but the variability itself.

But variability is just one of the larger lessons emerging from vom Saal's work. It has also opened a window on the powerful role of hormones in the development of *both* sexes and the extreme sensitivity of developing mammals to slight shifts in hormone levels in the womb. The wombmate studies have also underscored that hormones *permanently* "organize" or program cells, organs, the brain, and behavior *before birth*, in many ways setting the individual's course for an entire lifetime.

It is important to remember that hormones do this without altering genes or causing mutations. They control the "expression" of genes in the genetic blueprint an individual inherits from its parents. This relationship is similar to that between the keys on a player piano and the prepunched music roll that runs through and determines the tune. Though the piano can theoretically play many tunes, it will only play the one dictated by the pattern of holes in the music roll. During development, hormones present in the womb determine which genes will be expressed, or played, for a lifetime as well as the frequency of their expression. Nothing has been changed in the individual's genes, but if a particular note hasn't been punched into the music roll during development, it will remain forever mute. Genes may be the keyboard, but hormones present during development compose the tune.

What is astonishing about vom Saal's wombmate studies is how *little* it takes to dramatically change the tune. Hormones are exceptionally potent chemicals that operate at concentrations so low that they can be measured only by the most sensitive analytical methods. When considering hormones such as estradiol, the most potent estrogen, forget parts per million or parts per billion. The concentrations are typically parts per trillion, one thousand times *lower* than parts per billion. One can begin to imagine a quantity so infinitesimally small by thinking of a drop of gin in a train of tank cars full of tonic. One drop in 660 tank cars would be one part in a trillion; such a train would be six miles long.

The striking lifelong differences between a pretty sister and ugly sister stem from no more than a thirty-five parts per trillion difference in their exposure to estradiol and a one part per billion difference in testosterone. Using the gin and tonic analogy, the pretty sister's cocktail had 135 drops of gin in one thousand tank cars of tonic and the ugly sister's 100 drops—a difference that might not be detectable in a glass much less in a tank car flotilla.

This is a degree of sensitivity that approaches the unfathomable, a sensitivity, vom Saal says, "beyond people's wildest imagination." If such exquisite sensitivity provides rich opportunities for varied offspring from the same genetic stock, this same characteristic also makes the system vulnerable to serious disruption if something interferes with normal hormone levels—a frightening possibility that first dawned on vom Saal when Theo Colborn called him to talk about synthetic chemicals that could act like hormones.

To appreciate vom Saal's concern, one must understand more about the intricate choreography of events before birth known as sexual differentiation and the key role played by hormones in this developmental ballet. In mice, elephants, whales, humans, and all other mammals as well as in birds, reptiles, amphibians, and fish, the process that creates two sexes from initially unisex embryos is guided by these chemical messengers. They are the conductors that give the cues at the right moment as tissues and organs make now-or-never choices about the direction of development. In this central drama in which boys become boys and girls become girls, hormones have the starring role.

Our understanding of what determines whether a fertilized egg becomes a male or female is very recent. Before the twentieth century, it was widely assumed that the sex of the baby was determined by environmental factors such as temperature.

It was only in 1906 that two scientists—Nettie Marie Stevens and Edmund Beecher Wilson—independently noted that each cell in women had two X chromosomes while men always had an X and a Y, an observation that led to the theory that the number of X chromosomes determined sex. In the past decade, researchers have finally established that it is a gene on the Y chromosome rather than the number of X chromosomes that determines sex.

As most of us learned in high school biology, the eggs produced by the mother all carry one X chromosome, and the sperm from the father carry either an X or a Y chromosome. The sex of the baby hangs in the balance as the sperm burst out of the starting gate and race against each other in the reproductive marathon. If this most primordial of athletic events were broadcast like the Boston Marathon, we might hear that three Ys are neck-and-neck at the entrance to the cervix, but an X is making a move on the outside in the push into the uterus. A field of 75 million sperm have been pushing hard, sweeping their tails back and forth in steady swimming motions, but in the biological equivalent of Heartbreak Hill, many are beginning to flag as they enter the fallopian tube leading from the top of the uterus. It's a tight race right to the finish line as the competitors crowd toward the goal. At the finish line of this race, an egg awaits the victor, rather than a crown of laurel, as it crashes through. If the Y-carrying sperm gets to the egg first, the baby, who has XY chromosomes, will be a boy. If the first sperm to the egg carries an X, the XX chromosome will produce a girl.

Such stories about the race between the Xs and the Ys for the egg left many of us with the impression that the outcome was all in the genetic instructions carried by the sperm. If the sperm delivered a Y, bingo, it was a boy—what unfolded between conception and birth was all more or less automatic and dictated by that genetic blueprint. In fact, the process is much more complex. The sex-determining gene in that Y chromosome has only a quick walk-on part in the elegant and wondrous process through which boys become boys.

In animals such as birds and humans, one sex is the basic model and the other is what might be described as a custom job, since the latter requires a sequence of additional changes directed by hormones to develop properly into the opposite sex. In birds, this basic model happens to be male. In mammals, including humans, the opposite is the case, and an embryo will develop into a female unless male hormones override the program and set it off on the alternative course.

Although the sperm delivers the genetic trigger for a male when it penetrates the egg, the developing baby does not commit itself to one course or another for some time. Instead, it retains the potential to be either male or female for more than six weeks, developing a pair of unisex gonads that can become either testicles or ovaries and two separate sets of primitive plumbing—one the precursor to the male reproductive tract and the other the making of the fallopian tubes and uterus. These two duct systems, known as the Wolffian and Müllerian ducts, are the only part of the male and female reproductive systems that originate from different tissues. All the other essential equipment—which might seem dramatically different between the two sexes—develop from common tissue found in both boy and girl fetuses. Whether this tissue becomes the penis or the clitoris, the scrotal sack that carried the testicles or the folds of labial flesh around a woman's vagina, or something in between depends on the hormonal cues received during a baby's development.

The big moment for the Y chromosome comes around the seventh week of life, when a single gene on the chromosome directs the unisex sex glands to develop into male testicles. In doing this, the Y chromosome throws the switch initiating the very first step in male development, the development of the testes, and that is the beginning and end of its role in shaping a male. From this point on, the remainder of the process of masculinization is driven by hormone signals originating from the baby's brand-new testicles. In adult life, the testicles produce sperm to fertilize a woman's eggs, the male's contribution to reproduction and posterity. But the testicles play an even more important role in a male's life before birth. Without the right hormone cues at the right time—cues emanating from the testicles—the baby will not develop the male body and brain that go along with the testicles. It might not even develop the penis required to deliver the sperm the testicles produce.

In girls, the changes that turn the unisex glands into ovaries, the part of the female anatomy that produces eggs, begin somewhat later, in the third to fourth month of fetal life. During this same period, one set of ducts—the Wolffian ducts that provide the option for a male reproductive tract—wither and disappear without any special hormone instructions. While the development of the female body isn't as dependent on hormone cues as the development of males, animal research suggests estrogen is essential for proper development and normal functioning of the ovaries.

The process of laying the groundwork for the reproductive tract is more complicated in males and is marked by critical stages where hormones direct now-or-never decisions. Shortly after they are formed, the testicles produce a special hormone whose function is to trigger the disappearance of the female option—the Müllerian ducts. To accomplish this milestone, the hormone message must arrive at the right time, because there is only a short period when the female ducts respond to the signal to disappear. Then the testicles have to send another message to the Wolffian ducts, because they are programmed to disappear automatically by the fourteenth week unless they receive orders to the contrary.

The messenger is the predominantly male hormone testosterone, which insures the preservation and growth of the male Wolffian ducts. Under the influence of testosterone, these ducts form the epididymis, vas deferens, and seminal vesicles—the sperm delivery system that leads from the testicles to the penis.

A potent form of testosterone guides the development of the prostate gland and external genitals, directing the genital skin to form a penis and a scrotum that holds the testicles when they finally descend from the abdomen late in a baby's development. A naturally occurring defect dramatically illustrates what can happen if these messages do not get through.

From time to time, a young patient will show up in a gynecologist's office because the teenager still hasn't had her first period although all the other girls in her class have passed this milestone. Usually nothing serious is wrong.

But once in a rare while, the physician will deliver an utterly shocking diagnosis. The patient isn't menstruating because despite all appearances, she is *not* female. Although such individuals have grown up as normal-looking girls, they have the XY chromosomes of males and testicles in their abdomen instead of ovaries. But because a defect makes them insensitive to testosterone, they never responded to the hormone cues that trigger masculinization. They never developed the body and brain of a male.

The pictures in medical textbooks of these unrealized males are fascinating, for there is nothing about their unclothed bodies that looks the least bit odd or unusual. As hard as one searches for a hint that a genetic male lurks inside these bodies, there is no sign of development derailed. These genetic males look like perfectly ordinary women with normally developed breasts, narrow shoulders, and broader hips.

These completely feminized males are the most extreme example of what happens when something blocks the chemical messages that guide development. If anything interferes with the testosterone or the enzyme that amplifies its effect, the common tissue found in boy and girl fetuses will develop instead into a clitoris and other external female genitals. In less extreme cases of disruption, males may have ambiguous genitals or abnormally small penises and undescended testicles.

But sex is more than a purely physical matter. According to physicians who treat them, these feminized males not only look like women, they act and think of themselves as women. There is nothing the least bit telling in their behavior to suggest that they are really male. In most animals, the development of a properly functioning male or female involves the brain as much as the genitals, and research such as vom Saal's shows that hormones permanently shape some aspects of behavior before birth as much as they sculpt the penis. If an individual is going to act like a male as well as look like one, the brain must receive testosterone messages from the testicles during a critical period when brain cells are making some of their now-or-never decisions.

An individual who gets the wrong hormone messages during this critical period of brain development may show abnormal behavior and fail to mate even though it has the right physical equipment. In an influential 1959 study, Charles Phoenix of the University of Kansas found that female guinea pigs exposed to high levels of testosterone in the womb acted like males. They would not show the classic female mating posture, a raised posterior, known as "lordosis," as adults or respond normally to the female hormones that stimulate sexual behavior and reproduction.

No one debates that hormones act to give males and females different bodies and that their role in the development of animals and humans is pretty much the same. But how hormones influence the development of the human brain is hotly debated. Do they shape the brain and behavior in humans as dramatically as they do in mice or rats or guinea pigs? Are there structural differences between the brains of men and women, and is there any evidence that the differences stem from hormone influences before birth?

These questions are difficult to answer. Not only is human behavior more complex than that of vom Saal's mice, but we aren't free to give pregnant women various doses of hormones to see the effect on the brain development of their babies.

Those who have probed the question of whether the behavioral differences between men and women have a biological basis or are purely cultural have found evidence of some structural differences linked to hormones, but so far these sex-linked areas are fewer and less pronounced than those seen in rats. Psychologists have also re-ported certain general differences in the way men and women think, reporting that women have greater verbal skills as a rule and men tend to be better at solving spatial problems. Many also believe that the rough-and-tumble play and fighting seen to a much greater degree in young boys than in girls stems from biology rather than from culture or child-rearing methods.

At the same time that hormones are guiding at least some as-pects of sexual development of the unborn child, these chemical mes-sengers are also orchestrating the growth of the baby's nervous and immune systems, and programming organs and tissues such as the liver, blood, kidneys, and muscles, which function differently in men and women. Normal brain development, for example, depends on thy-roid hormones that cue and guide the development of nerves and their migration to the right area in this immensely complex organ.

For all these systems, normal development depends on getting the right hormone messages in the right amount to the right place at the right time. As this elaborate chemical ballet rushes forward at a dizzying pace, everything hinges on timing and proper cues. If something disrupts the cues during a critical period of development, it can have serious lifelong consequences for the offspring.

4

HORMONE HAVOC

As we pursue the mystery of hand-me-down poisons, two tragic episodes in medical history contain important lessons and immediate relevance to our quest. They leave no doubt that humans are vulnerable to hormone-disrupting synthetic chemicals and demonstrate that animal studies had repeatedly provided an early warning about the hazards for humans.

From the very beginning, these warnings were clear and ominous. As early as the 1930s, researchers at Northwestern University Medical School showed that tinkering with hormone levels during pregnancy was dangerous business, particularly for the fetus undergoing rapid development in the womb. In some of the experiments, the researchers simply gave an extra dose of estrogen to pregnant rats, who already have this female hormone in their bodies. The impact on their pups proved dramatic. At birth, the rat offspring showed striking abnormalities stemming from disrupted sexual development. The female pups exposed to extra natural or synthetic estrogen in the womb suffered structural defects of the uterus, vagina, and ovaries; males had stunted penises and other genital deformities.

In contrast to Fred vom Saal's work, which explores the effect of tiny *natural* variations in hormone levels in the womb, these earlier experiments boosted a female's hormone levels beyond the normal range by adding estrogen from outside the body. They showed that larger shifts in hormone levels scrambled the chemical messages and derailed sexual development. Although estrogen at normal levels is essential for development, too much of it can wreak havoc.

This cautionary evidence was timely indeed. In 1938, British scientist and physician Edward Charles Dodds and his colleagues had announced the synthesis of a chemical that somehow acted in the body like natural estrogen, and the medical community was abuzz with excitement. Leading researchers and gynecologists hailed the man-made estrogen, known as diethylstilbestrol or DES, as a wonder drug with a host of potential uses. Almost immediately, researchers began giving DES to women experiencing problems during pregnancy in the belief that insufficient estrogen levels caused miscarriages and premature births. What would prove to be a massive human experiment—one that eventually involved an estimated five million pregnant women in the United States, Latin America, and elsewhere—was just getting under way.

In the decades that followed, doctors not only prescribed DES to prevent miscarriages, they began to recommend it for untroubled pregnancies as if it were a vitamin that could improve on nature. Prestigious publications, among them the *Journal of Obstetrics and Gynecology*, carried drug company ads such as one from Grant Chemical Company that appeared in June 1957, which touted the use of DES for "ALL pregnancies," boasting that it produced "bigger and stronger babies."

DES also found a broad market beyond pregnant women. Doctors used it liberally to suppress milk production after childbirth, to alleviate hot flashes and other menopausal symptoms, and to treat acne, prostate cancer, gonorrhea in children, and even to stunt the growth in teenage girls who were becoming unfashionably tall. For years, college clinics doled out DES as a "morning after" contraceptive. Farmers were equally bullish on DES and used tons of it as an additive to animal feed or in neck or ear implants because it speeded the fattening of chickens, cows, and other livestock.

The postwar era was a time of Promethean optimism, when everyone from physicians to farmers rushed to embrace new "miracle" technologies. DES was just one of many new synthetic chemicals that promised to give us control over the forces of nature. With a mixture of hubris and naiveté, advocates of progress imagined a world with unlimited potential for the mastery of life itself.

The Northwestern University rat studies, which cast a dark shadow on the bold new era of hormone therapy, caused nary a ripple on the tidal wave of enthusiasm. Those who did take note of the findings tended to dismiss them as irrelevant to humans. The hormone-induced sexual abnormalities in rat pups were seen as a curiosity, as something that could happen only in rodents. Such skepticism is not unusual among physicians, whose anthropocentric tradition reinforces the notion that humans are a unique branch on the tree of life. Given this view, they were and still are inclined to regard human epidemiological studies as the only compelling evidence.

Medicine had also been dominated for decades by the myth of the placental barrier—the belief that the placenta, the complex body of tissue that attaches to the wall of the womb and to the baby through the umbilical cord, acts as an impenetrable shield protecting the developing baby from harmful outside influences. This myth endured long after evidence mounted to the contrary. According to the thinking of the time, the only thing capable of invading the womb and causing deformities was radiation.

In the next quarter century, two medical scandals would finally shatter this myth and radically redefine our ideas about the vulnerability of the baby during its time in the womb. The first bombshell was the thalidomide tragedy, which came to light in 1962, followed less than a decade later by the shocking discoveries about DES, a drug that doctors had been giving to women for more than thirty years.

When the story of the thalidomide babies broke, it caused an international sensation. Newspapers and magazines rushed to recount how a prescription drug had caused appalling birth defects. The pictures of infants without arms or legs touched a nerve, for it was every parent's worst nightmare.

Before doctors in Europe and Australia linked thalidomide to an alarming increase in bizarrely deformed babies, thousands of pregnant women had taken the drug as a tranquilizer or as a treatment for nausea. By the time it was finally removed from the market and medicine cabinets, it had caused severe deformities in eight thousand children in forty-six countries. The placenta had proved no barrier to the drug at all. The tragic episode also drove home the lesson that substances and doses tolerated readily by adults can devastate the unborn.

As would later be the case with DES, doctors came to suspect that something was seriously awry only because thalidomide caused abnormalities that were striking and almost unprecedented. Some babies were born with hands sprouting directly from their shoulders and no arms at all. Others lacked legs or were no more than a totally limbless trunk. The medical textbooks called the condition *pho-comelia*, which derives from Greek words for "seal" and "limb," because the hands or feet grow directly from the main joint like the flippers of a seal, but this birth defect was so rare before 1961 that the texts contained no photos. The malformation was illustrated for future physicians in one textbook with a drawing of a baby by the early nineteenth-century Spanish painter, Francisco de Goya.

Many children exposed to thalidomide before birth, however, showed no limb deformities but suffered instead from more commonly seen defects such as heart and organ malformation, brain damage, deafness, blindness, autism, and epilepsy. And some lucky children seemed to escape ill effects altogether although their mothers, too, had taken the drug during pregnancy. Why had some children been spared?

It wasn't that some mothers had taken a great deal of thalidomide and others had taken very little. Researchers found that the difference between devastating defects and an unscathed outcome appeared to depend on the *timing* of drug use, not on the dose. Possible genetic differences in sensitivity to thalidomide may have also played a role, making some individuals particularly vulnerable to the drug. Some of the mothers with limbless children had taken only two or three sleeping pills containing thalidomide during their entire pregnancy, but they had swallowed them at a critical period for the development of their babies' arms and legs—between the fifth and eighth week of pregnancy. The principle that "timing is all" would be demonstrated again and again as scientists explored the power of chemicals to disrupt development. A small dose of a drug or hormone that might have no effect at one point in a baby's development, for example, might be devastating just a few weeks earlier.

Americans largely escaped this tragedy, thanks to a skeptical physician at the Food and Drug Administration, Frances Kelsey, who had demanded more safety data and held up its general sale. Nevertheless, the thalidomide experience had a profound impact on the public and on scientists in the United States as well as elsewhere. Though it had taken an incident as subtle as a sledgehammer to drive the point home, the medical and scientific community finally accepted without question what some animal researchers had been trying to tell them for decades: chemicals can cause birth defects in humans as well as in rodents.

For ordinary people, the pictures of babies without limbs shook the technological optimism that had reigned since the end of World War II like a California earthquake and caused growing skepticism about the "wonder" drugs and "miracle" chemicals pouring onto the market and about the adequacy of government regulation. In the summer of 1962, through the news magazines such as *Life*, many shared the nightmare of Sherri Finkbine, a twenty-four-year-old mother and television host from Arizona who had taken thalidomide tranquilizers brought home from England during the critical early phase of her pregnancy. Convinced that her unborn baby had been severely damaged, Finkbine and her husband searched for a place in the United States to obtain an abortion, which was then illegal except to save the life of the mother. Their desperate quest finally ended in Sweden.

By chance, *Silent Spring*—Rachel Carson's now classic book on the dangers posed by synthetic pesticides to humans and the ecosystem—began appearing in serial form in *The New Yorker* just before the thalidomide story broke. The book caught the cresting wave of public anxiety and rode it to the best-seller list.

If thalidomide exploded the myth of the inviolable womb forever, the DES experience toppled the notion that birth defects have to be immediate and visible to be important.

* * *

Every parent prays for a normal, healthy baby.

When their daughter Andrea was finally born in September 1953, Eva and David Schwartz, a couple living in Boston's Roxbury neighborhood, felt they had been blessed with even more than they had prayed for. Their baby was not only normal and healthy, she was beautiful. Eva, who had suffered two miscarriages after the birth of her son Michael eight years earlier, was ecstatic. She often declared that the baby girl was the most gorgeous thing she had ever seen. Chubby, pink, blond, Andrea was the kind of baby you saw in baby food ads.

In a baby picture, Andrea's inquisitive eyes peek out from under the brim of a sunbonnet, suggesting intelligence as well as beauty. The face is framed by a delicate lace collar, and her mother's pride is evident in the impeccably ironed dress. As she grew, the little girl was tough and robust and never seemed to be anything but "perfectly healthy."

Then in April 1971, the Schwartzes' lives suddenly changed forever.

Andrea, then seventeen, was a high school senior, bubbling over with plans and dreams that started with college in the fall and eventually included marriage and a family. She had always wanted children, always. She remembers being captivated when she was just five, by her cousin's new baby and the thrill of being allowed to sit on the couch and hold it. Andrea didn't have overly grand ambitions; she just wanted a happy, normal life.

As Eva Schwartz paged through the *Boston Globe* one morning, she began reading a story that took her breath away. According to a new study in the *New England Journal of Medicine*, doctors at Massachusetts General Hospital had linked a rare vaginal cancer showing up in young women to a drug their mothers had taken during pregnancy—the synthetic estrogen DES. Her mind flashed on the pills, *hundreds* of pills, she had taken religiously while pregnant with Andrea. She had never missed a day, even though she often suffered from extreme nausea. On such days, she would wait until her roiling stomach settled a bit before taking the pills the doctor had ordered. Even before her medical records confirmed it, Eva Schwartz *knew*. She was one of those mothers.

Although that last pregnancy was an untroubled one, her physician at the neighborhood health clinic in Roxbury had nevertheless prescribed a regimen of DES, no doubt because of her history of miscarriages. Eva began taking the DES pills when she was just six weeks pregnant and continued to take increasingly larger doses of the drug as her pregnancy advanced, following a program recommended by a husband-and-wife research team from Harvard Medical School, George van Siclen Smith, a physician, and Olive Watkins Smith, an endocrinologist. The Smiths were the nation's leading advocates for prescribing DES for pregnant women, especially for those who had already lost pregnancies.

Andrea Schwartz would escape the worst ravages of DES. Unlike the most unfortunate ones, she did not die of cancer while still in her teens or undergo mutilating surgery that excised the uterus and vagina in the process of trying to excise the cancer. But medical tests over the next decade would show that despite outward appearances, Andrea was far from normal. DES had already stolen some of her dreams.

As one explores the DES experience, one nagging question recurs. It is a question that arises not just with DES, but with all hand-me-down poisons that cause damage across generations.

Would doctors have *ever* linked the medical problems suffered by young women with a drug their mothers had taken decades earlier if it hadn't been for a striking cluster of extremely rare cancers and a chance question posed by a patient's mother? Some specialists are confident they would have recognized the problem, saying that DES exposure causes other unique symptoms besides cancer, such as malformed vaginal tissue. Sooner or later, someone would have made the connection. Still, it is possible no one would have ever figured out that DES was doing profound but invisible damage to those exposed in the womb. Until DES, most scientists thought a drug was safe unless it caused immediate and obvious malformations. They found it hard to believe that something could have a serious long-term impact without causing any outwardly visible birth defects.

And even when one recognizes that prenatal events can lead to medical problems years later, the long lag time between cause and effect makes it difficult to prove connections or even verify that the mother had been exposed to the suspected drug or substance. In the case of DES, fears of liability on the part of doctors have only added to the difficulties for those exposed to the synthetic estrogen. DES daughters and sons joke bitterly about the epidemic of fires and floods that they say hit doctors' offices when they have sought to obtain their mothers' medical records.

The most painful aspect of the DES tragedy is that the drug did not even prevent miscarriages. By 1952, at least four separate studies had reported that women treated with DES for threatened miscarriages did no better than those treated with alternatives such as bed rest or sedatives. Later that same year, Dr. William Dieckmann and his colleagues from the University of Chicago presented a damning indictment of DES's efficacy at the annual meeting of the American Gynecological Society. In the largest and most carefully designed trial to date, the team enlisted two thousand pregnant women, treating half of them with DES and the rest with an identical-looking pill that contained no drug. The researchers attempted to eliminate bias by using double-blind methods that keep both patients and doctors ignorant of who got DES and who got the fake pill. Their conclusions were unambiguous: this wonder drug made no difference at all in the outcomes of pregnancies. Those who took DES did *not* have fewer miscarriages, fewer premature babies, or fewer infant deaths. Even worse, a later analysis of this same data concluded that DES had caused significant *increases* in miscarriages, premature births, and deaths among newborn infants.

Despite the studies showing DES to be ineffective, the federal Food and Drug Administration took no action to restrict its use during pregnancy. The University of Chicago study did, however, dampen enthusiasm somewhat, and some physicians stopped using DES. But many did not, and for almost two decades, hundreds of thousands of women still took DES during pregnancy, hoping to prevent miscarriages.

When the cluster of cancer cases began showing up at Massachusetts General Hospital in Boston, doctors were alarmed and utterly baffled. Between 1966 and 1969, specialists there had seen seven cases of clear-cell cancer of the vagina—an extremely rare cancer that almost never occurred in women under fifty. But the patients referred to the Harvard University teaching hospital for treatment during this period were all young women between fifteen and twenty-two. Prior to this rash of patients in one Boston hospital, only four cases had ever been reported in women under thirty in the world's medical literature.

Even the most radical treatment, which included removing the uterus and vagina, didn't always spare the young women's lives. One of those first patients died in 1968 at eighteen.

At first, Dr. Howard Ulfelder, a professor of gynecology at Harvard Medical School, had shrugged off the question posed by the mother of one of these young patients. She had taken DES during her pregnancy, she said. Did he think that could have anything to do with her daughter's illness?

Ulfelder just couldn't see how this could be possible. Nevertheless, when the next mother came in with a daughter suffering from clear-cell cancer, he decided to ask whether she had taken DES during pregnancy. He was stunned when she said yes.

Ulfelder and his Mass General colleagues Arthur Herbst, an obstetrician, and David Poskanzer, a medical epidemiologist, had been sifting through the histories of these young patients looking for some common factor that might explain why this rare cancer was suddenly showing up in these young women. At last, they had a possible clue. In April 22, 1971, they published a paper in the *New England Journal of Medicine* reporting that seven of the eight young women treated for clear-cell cancer of the vagina had mothers who had taken DES during the first three months of pregnancy.

Eva Schwartz couldn't bring herself to tell Andrea about DES and the danger of cancer for five months. Then just before Andrea was going to start college in the fall, she made an appointment to have her daughter examined by a gynecologist and broke the news.

Sitting at her kitchen table in the Boston suburb of Canton almost a quarter century later, Andrea Schwartz Goldstein cannot recall the words of that conversation. She only remembers the feelings, the sensation of being ripped from a sunny shore and sucked into a violent whirlpool of fear and uncertainty. Her mind raced. She thought about what she had confidently assumed about the life that stretched ahead of her. Maybe she would die of cancer before she ever had a chance to marry. Maybe she would never have the chance to have kids. Again and again after that day, the thought returned: "I'm not going to live a whole life."

She was still wrestling with the specter of cancer four years later, when she married Paul Goldstein, whom she had met through a friend when she was sixteen. The following year they bought a three-bedroom house in Canton, and Andrea took out a life insurance policy that was renewable without a physical, just in case. The insurance policy would help out if she died and left Paul with little kids.

At forty, Andrea is still blond and the green eyes still have the same intensity as in the baby pictures. She is an attractive woman who bears deep psychological as well as physical scars from her DES legacy. At times, she has battled depression, and even after all these years, the pain hovers just beneath the surface as she speaks, raw and unhealed. After thirteen years of working as an assistant to a physician specializing in infertility, she recently returned to school to get a nursing degree. DES took many things from her, she reflected, including the time of her life that should have been carefree and fun.

Although there were plenty of indications in animal studies that prenatal exposure to DES or estrogens might cause other damage, medical specialists focused for the most part on clear-cell cancer of the vagina and on tissue abnormalities in the vagina, which they feared might lead to cancer. It never occurred to Andrea that invisible damage caused by DES might make it impossible to have children. Even her doctors did not suggest the possibility.

Andrea puts another picture on the kitchen table not far from the baby picture of the blond girl in the sunbonnet. This picture, however, reveals the reality hidden beneath the appearance of health and normalcy. She holds the X ray up to the light, recalling the day she picked it up at the radiologist's office. The secretary handed it over, reporting that the doctor had said it was "the funniest uterus he had ever seen."

The previous year Andrea had suffered an ectopic pregnancy, an abnormal event in which the fertilized egg fails to descend properly into the uterus. Instead, it begins developing in the fallopian tube that leads from the ovary to the uterus—a dangerous situation that often ruptures the tube, causing severe bleeding and sometimes death. After Andrea was rushed to the hospital, doctors had operated, stopped the bleeding, and cut out the damaged tube. That left her with only one fallopian tube.

In the months that followed, she and Paul tried again to start a family, but to no avail, so she finally consulted a fertility specialist. When he learned that she was a DES daughter, her doctor had ordered a special X ray of the uterus because a new study had found that forty out of sixty daughters examined using a special dye and X-ray technique had abnormally formed uteri.

The eye searches the patterns of light and dark on the X ray looking for the familiar triangular space, the upside down pear seen so often in schematic diagrams of the female reproductive tract. There is nothing recognizable as a uterus. The area highlighted by the dye is a narrow, ragged tube containing no cavity at all.

Andrea liked her doctor because he didn't sugarcoat the diagnosis. He put it to her straight. Her uterus was "severely misshapen." She was never going to be able to have a baby.

Just as it had done to the rat pups in the experiments decades earlier, DES had left her with severe reproductive tract deformities.

Regardless of the compelling association between DES and vaginal cancer reported by the team from Mass General, some in the medical and scientific communities remained skeptical that DES could really cause vaginal cancer in those exposed before birth, even though animal studies a decade earlier had signaled possible links between early estrogen exposure and later cancers. In a 1963 study published in the *Journal of the American Cancer Institute*, Thelma Dunn, a pathologist with the National Cancer Institute, found that a variety of pathological changes, including cysts and cancers, developed in mice that had received estrogen injections as newborns. She warned that the results showed "the vulnerability of the immature animal to the harmful effects of exposure to a naturally occurring hormone." Dunn urged that when clues to the cause of cancer are sought in human populations, "every effort should be made to obtain the prenatal and early postnatal history of patients with cancer." A year later in the same journal, Noboru Takasugi and Howard Bern reported parallel findings, including permanent changes in the vaginal tissue of mice treated with estrogen shortly after birth. They, too, warned about the serious implications: "We feel that abnormal hormonal environments during early postnatal (and antenatal) life should not be underestimated as to their possible contribution to abnormal changes of neoplastic [cancerous] significance later in life." Although these animal studies proved to hold clinically relevant warnings for humans, doctors and drug makers paid no heed.

The argument about whether DES had caused the rare vaginal cancers was still raging in the early '70s when John McLachlan, a young researcher who specialized in the transfer of drugs and other environmental chemicals into the uterus, arrived at the National Institute of Environmental Health Sciences at Research Triangle Park, North Carolina, to set up a new group to investigate substances that disrupted development. The DES cancer question was one of the first challenges the developmental toxicology group took on.

Before long, the group had its first important finding. The vaginal cancer, which was rare in humans, had never been seen in mice. Nevertheless, McLachlan's team was able to induce adenocarcinoma of the vagina in female mice by giving DES to their pregnant mothers. The study helped settle that argument at least, but it was just the beginning of a still continuing debate about whether DES is responsible for a variety of medical problems and abnormalities seen in DES daughters and sons.

McLachlan and his colleagues then began to explore the effects of DES on male offspring. They demonstrated clearly that male mice exposed to DES in the womb were no less damaged than their sisters by this synthetic estrogen. These males had a variety of genital defects, including undescended testicles, stunted testicles, and cysts in the epididymis, a portion of the reproductive tract adjacent to the testicles where sperm mature. They also had abnormal sperm, reduced fertility, and genital tumors. The researchers found signs that DES had somehow interfered with the hormone messages during development. In normal male development, the testicles in the developing fetus produce a chemical messenger that triggers the disappearance of the female option, the Müllerian duct. The male mice exposed to DES still had parts of the female reproductive system. In 1975, the team published a major paper in the journal *Science* detailing the damage done to male mice exposed to this synthetic estrogen before birth.

Those were heady times, McLachlan recalls, full of excitement and discovery. It was cutting-edge science, and he loved being out there. His team kept in close touch with Dr. Arthur "Hap" Haney, a physician at the Duke University Medical Center in nearby Durham, who was treating humans exposed to DES. Time and time again, McLachlan would find something in a mouse and discover, when he called Haney, that the physician had seen the same problem in humans as well. Once in a while, the mouse findings would signal problems long before they emerged in humans. The developmental toxicology team warned about the possibility of undescended testicles three years before the problem was reported in boys whose mothers had been exposed to DES.

No doubt because they never developed a striking cancer, DES sons have been studied far less than DES daughters. While McLachlan and Haney repeatedly saw parallels between the damage found in mouse studies and the problems showing up in humans, this dearth of extensive human studies has made it impossible to establish conclusively that DES sons do indeed have such problems more frequently than those who were not exposed to estrogen prenatally. Some larger studies, such as those done at the University of Chicago, have reported higher rates of underdeveloped testes, stunted penises, and undescended testicles as well as a greater frequency of abnormal sperm in DES sons, but other studies haven't confirmed all these findings. Conflicting results have also emerged from studies exploring links between DES exposure and testicular cancer, although researchers have anticipated such a connection based on animal studies. In the eyes of the medical establishment, the question remains officially unresolved.

Despite the medical skepticism, many DES sons are convinced that they are suffering from DES damage, including greater rates of testicular cancer and fertility problems. Rick Friedman is one of those who is certain that DES has permanently marked his life.

For Friedman, youth hadn't been a time of robust good health. But he had forged ahead, ignoring as much as possible the arthritis and physical problems that plagued him. In his twenties, he had married and taken over the management of the family business, an International House of Pancakes that his father had built almost two decades ago in Ardmore outside of Philadelphia. If it wasn't one thing, however, it seemed to be another. After unsuccessful attempts to start a family, Friedman and his wife, Sachi, consulted a fertility clinic in 1987. While his wife suffered from reproductive difficulties, tests showed that he was part of the problem, too: he had abnormal sperm and a low sperm count.

It had never occurred to him that his growing litany of ailments might somehow be connected—the chronic allergies, the strange arthritis that hit when he was seventeen, the undescended testicle, the epididymal cysts in his reproductive tract, and the infertility that made it unlikely that he and his wife would have children. And then in 1992, a freakish tumor. The discovery that his problems could stem from a single cause, from something that happened to him *before* he was born, was sheer accident. How unexpected that a common thread might tie together all this pain and grief.

If Friedman had suspected he might be at special risk, perhaps he wouldn't have ignored the fatigue and breathing problems for so long. Perhaps he would have taken himself to a doctor before the tumor had mushroomed into "a mass the size of a child's football." Those were the words the doctor used to break the news that he was suffering from something more than overwork. At thirty-one years of age, Friedman found himself in the intensive care unit fighting for his life against cancer and the severe lung damage that the tumor had caused. The doctors gave him only one chance in three of pulling through.

Nevertheless, he beat the odds. He survived the operation and endured four rounds of aggressive chemotherapy. It was during his long recuperation in early 1993, a time when he could do little more than rest on the couch and read, that Friedman happened to pick up *McCall's* magazine. With long hours to pass, he found himself reading anything and everything, and on this day, he started perusing an article about the health problems of men who were exposed in the womb to a drug taken by their mothers to prevent miscarriages.

"Damn, this sounds like me," Friedman thought as he began to read through the problems reported in males exposed to DES in the womb—abnormal sperm, immune system problems such as arthritis, and the classic symptoms found in DES sons: undescended testicles and epididymal cysts. He read on. The article featured several young men who had been stricken with testicular cancer, which they are convinced is yet another legacy of DES. Because of the dearth of studies on DES sons and the difficulty of determining whether pregnant women took the drug decades earlier, however, researchers haven't firmly established this connection. Although the football-size tumor had grown in his chest, not his testicle, the doctors had told Rick it was a germ-cell tumor related to testicular cancer.

Rick pursued his hunch. At first his mother couldn't recall taking anything called DES, but that isn't uncommon. In one study, researchers found that only twenty-nine percent of the women whose medical records indicated they had taken DES during pregnancy could recall whether they had taken the drug or not. Another eight percent firmly stated that they had *not*, although their medical records reported the contrary. Then as his mother searched her memory, she told him that she had had a miscarriage before conceiving him and had shown signs of threatening another early in her pregnancy with him. Yes, she remembered, it was just after his grandmother had passed away. Her doctor had given her a shot and then put her on pills, but she didn't know if they were DES.

Friedman decided to track down his mother's medical records and met with the same frustration experienced by cancer researchers at Memorial Sloan-Kettering Cancer Center, who have tried to link testicular cancer with prenatal DES exposure. His mother's doctor was now dead and the records long gone. The pharmacy where his mother had bought the pills had moved and dumped their old records. The hospital where he was born did eventually come up with records, but they didn't contain information on his mother's prenatal care.

But Friedman really didn't need records to confirm in his own mind that he was indeed a DES son. Regardless of the continuing debate in the medical community, the circumstantial evidence seemed overwhelming, in his view. As he explored medical surveys and human and animal studies on DES effects, he found that he didn't have just one or two of the problems reported in males exposed to DES. He had half a dozen. "I am a textbook case," he concluded.

* * *

In the case of DES daughters, on the other hand, there is ample evidence that DES has caused clear-cell cancer and deformities of the reproductive tract. Because of these structural abnormalities, DES daughters are much more likely to have ectopic pregnancies, miscarriages, and premature babies. Although the majority of DES daughters eventually bear a child, often only after repeated attempts, the odds are stacked against them, with two out of three pregnancies ending in failure.

As was the case with thalidomide, the timing of the DES exposure appears more important than the dose. Women whose mothers took DES after the twentieth week of pregnancy do not suffer from the reproductive tract deformities, while those exposed before the tenth week of pregnancy have a greater chance of developing vaginal or cervical cancer.

This issue of timing also adds an element of confusion for those investigating how DES has affected humans. Lumping together individuals exposed early and late in pregnancy may mask the magnitude of the drug's impact on those exposed during critical periods of development. Many studies have treated DES-exposed individuals as a single group without addressing the question of timing.

But animal studies indicate that DES acts on other parts of the developing fetus besides the reproductive tract, including the brain, the pituitary gland, the mammary glands in the breast, and the immune system, causing permanent changes there as well. Researchers have found evidence that prenatal and neonatal exposure to DES or other estrogens can sensitize the developing fetus to estrogens and perhaps make it more vulnerable later in life to certain cancers, such as those in the breast, uterus, and prostate, that have been linked to elevated estrogen exposure.

In immune system studies on mice, scientists have found that DES exposure before birth reduces the number of T-helper cells, which are sometimes described as the heart of the immune system because they coordinate the overall immune response by telling other immune cells when to come into play. The importance of the T-helper cells has been vividly demonstrated recently by the arrival of the AIDS virus, which knocks out these key cells, thereby making the body incapable of mounting a coordinated immune response.

The devastation of the T-helper cells allows all kinds of invaders from cancers to fungus to run amok, which is why AIDS patients typically battle one disease after another. DES also affects another important part of the body's defense system, the natural killer cells, which are thought to act as a tumor patrol, alerting the immune system to the presence of tumor cells and controlling the spread of these cells to other parts of the body—a process known as metastasis. Given this decreased tumor defense, it is not surprising that a number of studies have found that DES-exposed mice show an increased sensitivity to chemical carcinogens in adult life and develop more cancers as they age.

Although DES-exposed mice have been reported to develop cancers in the breast, uterus, and ovaries, nothing is known about the incidence of these cancers in women exposed to DES before birth. But in the past decade, researchers have confirmed that women exposed to DES before birth show similar permanent changes in the function of their T and natural killer cells. Despite these impairments, DES offspring have generally not shown a greater vulnerability to infection, although one study did find an increased incidence of rheumatic fever. There is growing evidence, however, that DES-exposed women have a greater likelihood of developing autoimmune diseases, such as Hashimoto's thyroiditis, Graves' disease, rheumatoid arthritis, and other diseases stemming from defects in the regulation of the immune system. Based on animal studies that show that the severity of immune defects increases with age, researchers are concerned that humans exposed to DES will also experience more immune system problems as they get older.

Has DES had as dramatic an impact on the brain as it has had on the body? This is perhaps the most intriguing question stemming from the inadvertent human DES experiment, and it remains largely unanswered. Although studies have found that DES disrupts the physical development of humans and animals in identical ways, it isn't clear that these striking parallels hold when it comes to its impact on the development of the brain. Studies have found some differences between species in the balance of hormones at work during this process, making it difficult to extrapolate directly from rodents to humans.

In animals, exposure to DES or higher than normal levels of estrogen causes "dramatic and permanent changes in brain structure and behavior," according to Melissa Hines, a researcher at Goldsmiths' College, University of London at New Cross, who specializes in the effect of hormones on the development of the brain and behavior. And here again the effects are surprising and counterintuitive. As strange as it might seem, female rats, mice, hamsters, and guinea pigs exposed to excess estrogen before or just after birth show a more *masculine* pattern of reproductive behavior. They mount other animals more frequently and are less inclined to display the female mating posture. Early estrogen exposure also alters other behavior that differs between the sexes, such as rough-and-tumble play, maze learning, and aggression, causing females again to act more like males. So while small amounts of estrogen appear necessary for normal female development, higher doses result in masculinization. Across a wide range of species, including amphibians, songbirds, rodents, dogs, cattle, sheep, and rhesus monkeys, researchers have found that exposing developing females to increased estrogens or male androgens will increase masculine behavior and decrease feminine behavior.

In male mice whose mothers are treated during pregnancy with low doses of DES, scientists have observed increased rates of territorial behavior, notably the marking of their territory with urine, and obvious increases in activity levels as adults. At high doses, one gets the opposite effects and impaired masculine behavior.

The animal studies raise provocative questions about possible effects in humans, but unfortunately most of them have never been carefully investigated, Hines notes.

The best evidence suggesting a link between DES exposure and human behavior comes from studies on sexual orientation, the question of which sex an individual finds sexually attractive. The vast majority of men are attracted to women and the vast majority of women to men, an evolutionarily unsurprising arrangement since it facilitates reproduction. But the classic Kinsey studies on human sexual behavior found that this difference is not absolute. The Kinsey surveys done in the late '40s and early '50s reported that about ten percent of the men were sexually attracted to other men and three to five percent of the women to other women.

Hines summarized several studies comparing DES-exposed women with their unexposed sisters or with other unexposed women and found an association between prenatal DES exposure and homosexuality and bisexuality. In one of these studies, researchers recruited and interviewed sixty women from a medical clinic—thirty who had been exposed to DES and thirty presumably unexposed women who showed abnormal Pap smears (although this condition is also associated with DES exposure). The researchers conducted interviews with these women and assessed their sexual orientation using a seven-point heterosexual to homosexual gradient developed by Kinsey. While none of the women with abnormal Pap smears indicated a homosexual or bisexual orientation, twenty-four percent of the DES women reported a lifelong bisexual or homosexual orientation. These researchers also compared twelve DES daughters with sisters who had not been exposed to the synthetic estrogen and found that forty-two percent of the DES-exposed women had a lifelong bisexual orientation, but only eight percent of their sisters. A second study of thirty DES-exposed women and thirty unexposed controls who were matched based on such characteristics as age, race, and social class reported similar differences. Studies on DES sons so far have found no indication that DES influences the sexual orientation of sons.

The research team that recruited women at the medical clinic also interviewed them to see if they could detect differences in other behaviors that typically vary between male and females, such as parenting interests, degree of physical activity, and aggression and delinquency. An initial study reported that those exposed to DES showed less interest in parenting, but a second failed to find such an association with DES exposure.

Researchers have also found surprisingly high rates of major depression in both men and women exposed to DES before birth as well as other psychiatric disorders such as anxiety, anorexia nervosa (an eating disorder in which individuals starve themselves by refusing to eat), and phobic neurosis. Studies have reported such differences even when DES sons and daughters were unaware that they had been exposed. In one series of studies of the DES-exposed, forty percent of the women and seventy-one percent of the men had experienced major depression that impaired the individuals' ability to function at home, work, or school, required taking medication for depression, or required professional help.

The DES experience is rich in lessons.

This tragic and unintended experiment demonstrated that chemicals could cross the placenta, disrupt the development of the baby, and have serious effects that might not be evident until decades later. This was a previously unrecognized medical phenomenon: *delayed* long-term effects that did not emerge until the child reached puberty or anytime later in life.

It warned that appearances are not always coincident with reality. One had to worry not just about gross and immediately apparent birth defects such as missing limbs but about invisible damage done during development to tissue and cells—damage that can, nevertheless, have a lifelong impact and undermine survival.

It dramatized the dangers of interfering with the delicate balance of hormones during development. It showed how fragile the fetus is and how it passes through critical stages when it is particularly vulnerable. It underscored that an unborn baby is not just a small adult. Drugs and chemicals that have little effect on adults can cause serious and permanent damage to a baby during its rapid prenatal development.

Again and again, the DES experience brought home the common fate of mice and men. Rodents and humans exposed to DES in the womb suffer identical damage to the genitals and the reproductive tract, a parallel that also holds true not just for mammals but for many other animal types as well. To an astonishing degree, evolution has conserved through hundreds of millions of years a basic strategy in vertebrates for embryonic development that is dependent on hormones. Regardless of whether the offspring is a human or a deer mouse, a whale or a bat, hormones regulate its development in fundamentally the same way.

"There are so many apparent differences in these species," John McLachlan noted. "Yet the strategy for sexual development is remarkably similar, and the effect of estrogen is remarkably similar. That sounds simple," he reflected, "but to my thinking it's profound."

* * *

The DES experience offered another critical lesson as well that is relevant not just to those exposed to DES but to all of us. The developmental effects of DES made it clear that the human body could mistake a man-made chemical for a hormone. In the mid-1970s, researchers began to discover that other man-made chemicals, such as the pesticides DDT and kepone, showed hormone effects as well. It would take time for McLachlan and others to recognize the potential importance of this observation.

5

FIFTY WAYS TO LOSE YOUR FERTILITY

DESPITE THE PROBLEMS IT WOULD LATER CAUSE, DES LIVED UP TO THE promise in one regard—it mimicked natural estrogen. This fact alone is an intriguing puzzle, for this man-made chemical bears surprisingly little structural resemblance to natural estrogen. How could it act like a hormone?

This question lies at the heart of the deepening mystery of how foreign chemicals trick the body and disrupt its own chemical messengers. In the past half century since DES appeared, scientists have learned that DES is not unique in its hormone effects. One by one, they have stumbled upon many other chemicals—both man-made and natural compounds—that act like hormones, and gradually the realization has dawned that the world is full of hormone disruptors. Unlike DES, however, most don't come in little pills.

By an interesting coincidence, in the very same year that Edward Dodds announced the synthesis of DES, a Swiss chemist, Paul Müller, discovered a powerful new pesticide, and both synthetic chemicals made their debut amid great acclaim in 1938. Just as DES was heralded as a "wonder drug," DDT was hailed as a "miraculous pesticide."

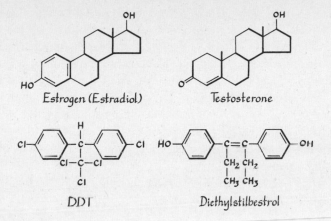

Estrogen (Estradiol)   Testosterone

DDT   Diethylstilbestrol

Chemical structures of the natural female hormone (estradiol), generically called "estrogen," and the natural male hormone, testosterone, compared with the structures for DDT, an insecticide, and diethylstilbestrol (DES), a synthetic female hormone drug with estrogen-like effects. Surprisingly, DDT and DES proved to be effective mimics of natural female hormones, despite their dissimilar appearance.

Dodds received a knighthood for his efforts in synthesizing sex hormones, and Müller won the Nobel Prize in 1948.

Twelve years after the advent of these compounds, researchers at Syracuse University learned the two chemicals shared a deeper kinship. Although DDT had been developed to kill insects and not for use as a drug or synthetic hormone, it, too, seemed to have the effect of estrogen when it was given to young roosters: it feminized them. The males treated with DDT had severely underdeveloped testes and failed to grow the ample combs and wattles that roosters display. In considering these results, Verlus Frank Lindeman and his graduate student Howard Burlington noted that the chemical structure of DDT bears a similarity to that of DES.

However much these two synthetic chemicals resemble each other, these impostors do not look much like estrogen or the other steroid hormones made by the body itself. The steroid hormone family is one of three hormone groups, which are generally classified according to their chemical structure or function. The steroid hormones that help carry on the body's never-ending internal conversation all share a common architecture based on four rings. The male and female hormones, testosterone and estrogen, may have powerfully different effects, but in diagrams of their chemical structure, they are remarkably similar. The divergent destinies of male and female hinge on an atom here and there. By contrast, DDT and DES have a two-ringed configuration. The difference between this arrangement and that of estrogen is immediately apparent even to someone who has never taken chemistry. Based on their structure, it would be impossible to mistake these synthetic chemicals for members of the steroid hormone family.

Yet for reasons that still aren't fully understood, the body does mistake them for the real thing.

The plastic model John McLachlan holds in his hand looks like a mass of colored bubble-gum balls. It is the size and general shape of a small loaf of Italian bread.

More than two decades after he first embarked on his exploration of DES, McLachlan is sitting on the edge of a table in his office at the National Institute of Environmental Health Sciences, giving a lesson in Chemical Messengers 101—the basics on how the body communicates through hormones. Like many natural teachers, he has a theatrical flair and a penchant for metaphor. He reaches automatically for a prop to demonstrate his point. This isn't simply science, it is a fascinating story—the tale of the estrogen receptor, which consorts so readily with foreigners that it has earned a reputation. Some scientists call it "promiscuous."

The plastic model is a gargantuan representation of estradiol, one of the three principal types of estrogen manufactured by the ovaries and dispersed into the bloodstream.

McLachlan, a fifty-year-old man with a head of curly gray locks and merry dark eyes that gleam like onyx, then cups his free hand. This is an estrogen receptor, a special protein found inside cells in many parts of the body, including the uterus, the breasts, the brain, and the liver. The receptor receives the chemical message, in this case estrogen, sent from the ovaries, picking up signals from the bloodstream in the same way a cellular phone picks up radio signals in the air. A receptor isn't supposed to receive all the chemical signals flying about. Like a cellular phone, it is supposed to receive only those intended for it.

The body has hundreds of different kinds of receptors, each one designed for a particular kind of chemical signal. Some receive messages from the thyroid gland, which may cue cells to consume more oxygen and generate more heat. Others are tuned to the adrenal glands, which send messages that regulate blood pressure and the body's response to stress. The hypothalamus in the brain has all kinds of receptors to monitor hormone levels in the blood so the brain can signal the hormone-producing glands when adjustments are needed. And there is a whole class of mystery receptors, known as "orphan" receptors, that are tuned to messages that scientists have not yet identified.

Each hormone and its particular receptor have a "made for each other" attraction, which scientists describe as a "high affinity." When they encounter one another, they grab hold, engaging in a molecular embrace known as "binding."

McLachlan demonstrates by moving the plastic model through the air toward the receptor, showing how the estradiol docks in the pocket of the receptor like a Star Trek vehicle returning to the much larger mother ship. Hormone molecules are tiny compared to the sprawling receptors.

They fit together, he notes, like a lock and key, and once joined, they move into the cell's nucleus to "turn on" the biological activity associated with the hormone. This union of hormone and receptor targets genes that trigger the production of particular proteins. In the case of estrogen, these proteins accelerate cell division. So when estrogen joins with receptors in the uterus, it will cause the lining of that organ to thicken. Estrogen produces such a response in the first half of the menstrual cycle to prepare the uterus in the event an egg is fertilized when ovulation occurs at midcycle.

This lock-and-key notion has dominated the theory of how the body communicates through hormones. In endocrinology textbooks, one still finds flat assertions that receptors are highly discriminating about chemical structure and will bind only to their intended hormone or a very closely related compound. Although theory holds true in a general way, reality is proving considerably messier and unpredictable,

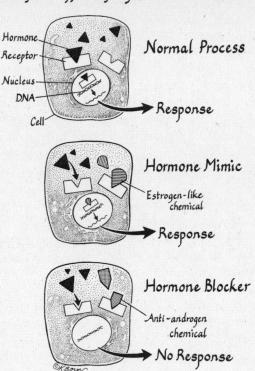

Receptor Effects of Synthetic Chemicals

Normal Process

Hormone
Receptor
Nucleus
DNA
Cell

Response

Hormone Mimic

Estrogen-like chemical

Response

Hormone Blocker

Anti-androgen chemical

No Response

Hormones and their receptors fit together with a "lock and key" mechanism. Under normal conditions (*top*), a natural hormone binds to its receptor and activates genes in the nucleus to produce the appropriate biological response. Hormone mimics (*middle*) can also bind to the receptor and induce a response. Hormone blockers (*bottom*) do not induce a response, but prevent natural hormones from attaching to the receptor. Certain synthetic chemicals released into the environment can behave like hormone mimics and hormone blockers, contributing to disruption of cellular activity. The compound that outnumbers or outcompetes for receptor sites determines the response by the cell.

not only in the case of the estrogen receptor but with other hormone receptors as well.

When Dodds and his colleagues announced they had developed a synthetic estrogen, they did not understand how DES was able to mimic the hormone in the body. They just knew empirically that it worked. A quarter century passed before other researchers discovered the receptors that receive the chemical messages and finally came to understand what made DES effective. It somehow insinuates itself into this estrogen receptor.

As he explains this, McLachlan maneuvers a model of a DES molecule into the pocket of his imaginary receptor, which readily accepts it as the real thing. Surprisingly, this chemical con artist triggers the system more effectively than estradiol, the body's own estrogen.

Perhaps even more important, research has discovered that this synthetic chemical manages to circumvent a mechanism that protects a developing fetus from excessive estrogen exposure, which can disrupt its development. The blood of the mother and the developing fetus contain special proteins that soak up almost all the estrogen circulating in the blood and make it unavailable to receptors. But these proteins—called sex steroid binding globulin—do not recognize DES and thus do not bind with it. As a consequence, only a tiny fraction of the natural estrogen in the bloodstream will be free, but *all* the DES will be biologically active. Whether these protective substances recognize and soak up other man-made hormone mimics is a major unanswered question, but evidence suggests that these, too, can make an end run around them—an unfortunate fact if true, for that leaves the unborn all the more vulnerable to disruption. Without this defense mechanism to prevent overexposure to estrogenic chemicals, even seemingly low concentrations of hormone mimics may still pose a hazard.

The relative strength of hormone impostors is another consideration. Most hormone impostors are considerably less potent than DES or estradiol because they do not bind as firmly to the estrogen receptor. Some scientists have therefore suggested that these "weak" estrogens are probably not powerful enough to cause problems. Howard Bern, a distinguished researcher who has explored the effects of weak estrogens, is not so sanguine. Bern is a comparative endocrinologist at the University of California at Berkeley and a major figure in experimental DES research.

"The real issue is the special sensitivity of the developing organism," Bern says. It may be particularly vulnerable not only because it is undergoing rapid development but also because its hormone receptors aren't as discriminating as those of an adult. "It may not see the difference between weak and strong estrogens."

In experiments with mice, Bern found that so-called weak estrogens seem to have a far more potent effect on the unborn than on exposed adults. What happens in adults, he stresses, is no basis for predicting what these chemicals can do to the unborn.

It is also important to keep in mind that natural estrogens operate at extremely low concentrations, measured in parts per trillion. In contrast, these so-called weak estrogens are present in human blood and body fat in concentrations of parts per billion or parts per million—levels sometimes thousands to millions of times greater than natural estrogens. So even though the contaminant levels may seem miniscule, they are not necessarily inconsequential.

The understanding of hormone receptors, which has grown rapidly since they were first identified in the mid-1960s, also sheds light on why DES and other hormone disruptors have such similar effects across an astonishing range of species. Classic accounts of evolution tend to emphasize innovation and change in the story of life on Earth, but there has been a strong conservative streak in evolution as well. A good deal has persisted through eons largely unaltered, especially elements of basic design, such as the endocrine system.

As scientists have explored hormone receptors in different animals, they have marveled at the *lack of change* over millions of years of evolution. Whether in a turtle, or mouse, or a human, the endocrine system produces a chemically identical estradiol that binds to an estrogen receptor. The discovery of similar estrogen receptors in animals as distinct as turtles and humans suggests that the internal communication system based on hormones and receptors is an ancient adaptation that arose early in the evolution of vertebrates—the evolutionary branch of animals with backbones that includes humans. Scientists believe that turtles have undergone little change since they arose from a reptilian ancestor over 200 million years ago, long before modern mammals appeared on the scene.

Although receptor research demonstrated that impostor chemicals such as DDT and DES do bind with the estrogen receptor, it has not really illuminated why the receptor readily accepts them. The similarity between DDT and DES led scientists to expect that they might find a common structural feature to explain the phenomenon, but the mystery of hormone mimics would not yield to such a simple explanation. To their bewilderment, they found that the estrogen receptor binds to chemicals with a variety of strikingly different structures. It is a lock that can be opened with devices that bear as little resemblance to natural estrogen as a hammer does to a key. Even more puzzling, a wrench might work as well as a hammer.

DDT was, moreover, just the first surprise. At roughly the same time that researchers in the United States were giving the pesticide to chickens, other scientists from a distant continent and an entirely different field would stumble upon another estrogen mimic in the most bizarre place.

The early 1940s seemed like a particularly promising time for the sheep ranchers in the gently rolling hills south of Perth in western Australia. Three unusually good seasons had followed one after another, and with the favorable weather, the pastures exploded into lush, green growth, allowing the sheep to graze for an exceptionally long time. According to the ranchers in the region, the sheep—handsome, burly merinos that produce fine, luxurious wool—had never looked so good.

But just when things had never been better, a strange epidemic began hitting the flocks—an epidemic of infertility. The first sign was a striking increase in stillborn lambs. Then the ewes carrying lambs failed to go into labor; the lambs died and often the mothers as well. Each year the problem worsened until finally, even after repeated breeding to fertile rams, most of the ewes simply did not conceive at all. In a matter of five years, the breeding programs stopped cold, and ranchers in the area faced financial disaster. Without the irrepressible exuberance of the gamboling lambs, spring did not really seem like spring.

After extensive detective work that involved not only the state agricultural specialists but federal scientists as well, researchers finally determined that the cause of the sterility epidemic wasn't to be found in poison or disease or a genetic defect. The cause was clover.

Fifteen years earlier, ranchers had started to work on improving their natural pastures by sowing a species of clover native to the

Mediterranean region in Europe. The early strain of subterranean clover seemed equally well suited to the local climate and in a short time it brought great increases in the productivity on these ranches. For reasons the researchers could not pinpoint at first, it also caused this strange reproductive malady, which they named "clover disease."

The first scientific paper on this phenomenon appeared in the *Australian Veterinary Journal* in 1946, but it took several more years to isolate three chemicals suspected of causing the sterility. In the end, however, researchers determined that only one of these chemicals, formononetin, was the culprit. This natural compound, which escapes breakdown in the sheep's stomach, can, like DES and DDT, mimic the biological effects of estrogen.

Surprisingly, plant evolution had produced chemicals that mimic estrogen long before Dodds synthesized DES in the laboratory, and not just one or two, but many—twenty of which are now known to science. To date, researchers have found these estrogenic substances in at least three hundred plants from more than sixteen different plant families. The list includes many foods that feed the world as well as some of our favorite herbs and seasonings. Hormone mimics lurk in parsley, sage, and garlic; in wheat, oats, rye, barley, rice, and soybeans; in potatoes, carrots, peas, beans, and alfalfa sprouts; in apples, cherries, plums, and pomegranates; and even in coffee and bourbon whiskey. Like DES and DDT, these plant compounds can fool the estrogen receptor.

If the clover in Australia were the only case of a natural hormone mimic in the annals of science, it might be shrugged off as an evolutionary fluke, but the presence of an estrogenic substance in so many diverse plant species suggests this is no accident.

So why are plants making estrogens?

"Plants are making oral contraceptives to defend themselves," says Claude Hughes, a researcher exploring the effects of hormone-like compounds on the reproductive system. It might sound like a wild idea, but from an evolutionary perspective it makes sense.

Since plants cannot escape predators by running away, they have evolved a fascinating variety of defenses. Some smell bad, taste bad, or poison those that eat them. Others have unpalatable thorns, spines, or indigestible substances in their leaves. When insects attack, many plants fight back with a chemical arsenal that can kill insects outright, make them stop feeding, or disrupt their growth by mimicking insect growth regulator hormones. This growth disruption typically makes an insect sterile, thus reducing the troubling insect population.

The more Hughes explored the notion that plants might be making contraceptives, the more evidence he found consistent with the theory that this is indeed what plants are up to. By lacing their leaves with hormonally active substances, they suppress the fertility of the animals that feed on them. By Hughes's theory, clover disease isn't simply an unfortunate livestock malady, it is a subtle and previously unrecognized form of plant self-defense. The plants that make estrogen mimics, he notes, are tasty ones sought out by animals and humans for food, not the unappetizing plants that contain foul-tasting compounds—an alternative defensive strategy.

Hughes is a specialist in reproductive endocrinology at the Bowman Gray School of Medicine, Wake Forest University, in Winston-Salem, North Carolina, who holds an M.D. as well as a Ph.D. in neuroendocrinology, the study of the interaction between the brain and hormones. As an undergraduate, he pursued research in plant physiology. He is also the son of a farmer and now raises sheep on his own farm in North Carolina, so he brings firsthand experience to the task as well.

The thought that plants might be making chemicals aimed at undermining the fertility of their predators first occurred to Hughes while he was a doctoral student studying the impacts of marijuana on the brain. Humans have long used marijuana as a drug because the chemicals it contains act in the brain to alter mood and perception, creating a "high." But as Hughes and others discovered, these chemicals do more than induce a pleasant mellowness; they interfere with reproduction in a variety of ways. The same compound that makes a pot smoker high also acts on the testicles to reduce the synthesis of testosterone and on the brain to suppress lutenizing hormone, a key hormone that cues ovulation in females and testosterone production in males. Studies have reported that marijuana feminized men who smoked it heavily.

Hughes's work focused on the way that marijuana interferes with the hormone prolactin, which is produced in the brain and signals the breast to produce milk. Mother rats given marijuana produced no milk, and their pups died of starvation. Hughes later moved on to investigate the effects of plant estrogens on the endocrine system and the hormones that orchestrate reproduction, an area few scientists had explored.

For such a defensive strategy to work, he explains, the plant would logically target females rather than males because a predator's reproduction is limited by the number of fertile females. If, for example, a plant managed to impair the fertility of all the males save one, that single male can, nevertheless, fertilize an entire flock of females. But if only a single female is fertile, she can produce only one or two lambs.

Plants containing estrogen mimics produce them according to a seasonal pattern that fits perfectly with this strategy. Clover packs the greatest concentrations of estrogenic compounds into the new growth in spring, and when a rabbit or a sheep injures it by munching on these tender shoots, the plant responds by producing *even more* estrogen at the site of injury, delivering an added dose to predators that continue grazing.

Humans long ago figured out that certain plants have contraceptive powers, judging from references in classical literature. Historian John M. Riddle of North Carolina State University reports that women throughout the ancient world used a variety of plants to prevent pregnancies and precipitate abortions, including a now extinct giant fennel called silphium. Researchers have confirmed that many plants in the fennel family produce estrogenic substances or other hormonally active compounds. The ancients also used wild carrot, the beautiful and delightfully common weed now known as Queen Anne's lace, which the Greek physician Hippocrates, who lived in the fourth century B.C., described as having similar powers. Studies have shown that its seeds contain chemicals that block the hormone progesterone, which is necessary for establishing and maintaining pregnancy.

The pomegranate played a central role in both Greek myth and their birth-control efforts. According to the myth, Persephone, the daughter of the fertility goddess Demeter, was told to eat nothing during a visit to the underworld Hades, but she disobeyed and ate a pomegranate. As punishment, the gods sentenced her to spend a part of the year in the underworld, and for this reason, Earth experiences the barren season of winter until Persephone returns each spring. Riddle says the Greeks used pomegranate as a contraceptive, and here again studies have found that it contains a plant estrogen that acts like the chemicals found in modern oral contraceptives manufactured by the pharmaceutical industry.

The presence of estrogenic compounds in so many foods raises an important question. Do these substances pose a hazard to human health or to the development of babies?

There is no simple answer to this question. Plants containing estrogen mimics may be beneficial in some instances and hazardous in others, according to Patricia Whitten, an anthropologist working at the Laboratory of Reproductive Ecology and Environmental Toxicology at Emory University in Atlanta, Georgia. Scientists are just beginning to explore plant estrogens and how these hormone mimics in food affect us, so fundamental questions— such as how much we actually ingest in our foods—remain as yet unanswered. Because humans eat a varied diet, it is not clear whether we ingest sufficient quantities to worry about. The dose question is, moreover, inherently tricky when dealing with hormones. Depending on your age, sex, and hormonal status, the same dose can have wildly different effects. It will matter whether you are a man or a woman; a postmenopausal woman or one still in her reproductive years; an adult, a child, or a baby developing in the womb.

Whitten has found that exposure to plant estrogens early in life can undermine the ability of rat pups to reproduce when they grow up. In her experiment, the rat mothers were given low doses of coumestrol, a plant estrogen found in sunflower seeds and oil and alfalfa sprouts, which they passed on to their babies through their milk. Rats are considerably less developed than humans at birth, so in the days after birth they are undergoing stages of development that in humans occur in the womb.

The pups in this experiment did not suffer obvious genital defects or other physical abnormalities in the reproductive tract as seen in the DES experiments, but they showed evidence of permanent changes that sabotaged their fertility.

"We think we've altered the sexual differentiation of the brain," Whitten says of the exposure. The females don't ovulate and are sterile because their brains do not respond to the hormone that triggers ovulation—an indication that they have been masculinized. The males, on the other hand, are feminized, showing less mounting behavior and fewer ejaculations. For a rat, the first ten days after birth are the critical period for the development of those areas of the brain linked to sexual behavior.

But the very same foods that disrupt development before birth or early in life might help prevent disease in an adult. Evidence that foods high in plant estrogens, such as soybeans, might protect against breast and prostate cancer has sparked a great deal of scientific interest and new research into plant estrogens. Numerous studies have linked estrogens, even those naturally occurring in the body, to cancer, suggesting that the greater a woman's lifetime exposure, the greater the risk. Researchers theorize that plant estrogens might be protective because they are weaker than the natural estrogens made in the body. If they occupy estrogen receptors in the breast and displace natural estradiol, they might reduce a woman's lifetime exposure to estrogen.

The thing to keep in mind, Hughes says, is that plants and the animals that eat them, including humans, share a long evolutionary history. Over many generations, the most sensitive individuals, those who became sterile from eating estrogenic foods, dropped out of the population. All those who were able to produce at least some offspring passed on a certain degree of resistance. This sort of evolutionary winnowing occurs because of individual differences.

The discovery that DDT could act like an estrogen must have seemed like a singular curiosity in 1950, but unfortunately, it has proved far from unique. Over the past half century, the same chemical laboratories that produced this "miraculous" pesticide created a host of other synthetic chemicals that can also interfere with hormones. We have been slow to recognize this threat or to realize that the world has become permeated with hormone-disrupting synthetic chemicals.

When male workers in a chemical plant developed extremely low sperm counts after exposure to the pesticide kepone, it became clear that DDT was not the only synthetic chemical capable of producing estrogenlike effects. Others were quickly added to the list. Like DDT, these synthetic chemicals were not intended as drugs or hormone mimics. They were invented by chemists in laboratories to kill insects threatening crops and to give manufacturers new materials such as plastics. Inadvertently, however, the chemical engineers had also created chemicals that jeopardize fertility and the unborn. Even worse, we have unknowingly spread them far and wide across the face of the Earth.

How many man-made chemicals scramble the body's chemical messages? No one knows and no one has systematically screened the tens of thousands of synthetic chemicals created since World War II for such effects. As with kepone, which is now banned, many of those that are known have been discovered by accident.

To date, researchers have identified at least fifty-one synthetic chemicals—many of them ubiquitous in the environment—that disrupt the endocrine system in one way or another. Some mimic estrogen like DES, but others interfere with other parts of the system, such as testosterone and thyroid metabolism. This tally of hormone disruptors includes large chemical families such as the 209 compounds classified as PCBs, the 75 dioxins, and the 135 furans, which have a myriad of documented disruptive effects.

Most discussions of hormone-disrupting chemicals inevitably focus on DDT, the PCBs, and dioxin, but not because they necessarily pose the only or the gravest threat. These get the lion's share of the attention because they happen to be the only hormone-disrupting chemicals that scientists have studied in any depth. While admittedly far from the whole story, these well-known cases do, however, serve to illustrate a much broader problem, so they will also receive considerable attention in this book. The magnitude of this problem is still unclear, but those who have watched the list of hormone disruptors grow think the age of discovery is far from over. "There are probably a lot more," says John McLachlan.

As the number of hormone-disrupting chemicals mounts, Claude Hughes worries, emphasizing that humans lack evolutionary history with these synthetic compounds. These man-made estrogen mimics differ in fundamental ways from plant estrogens, he notes. The body is able to break down and excrete the natural estrogen mimics, while many of the man-made compounds resist normal breakdown and accumulate in the body, exposing humans and animals to low-level but long-term exposure. This pattern of chronic hormone exposure is unprecedented in our evolutionary experience, and adapting to this new hazard is a matter of millennia not decades. He worries that some portion of the population is bound to be sensitive. He worries about his daughter and son and the grandchildren he anticipates in years hence. What if his kids are among the sensitive? What if they can't reproduce because of eating this stuff?

Some might be tempted to jump to the conclusion that because so many natural estrogens already exist in nature, there is therefore no need to worry about synthetic chemicals that interfere with hormones. This kind of argument has surfaced in the discussion about carcinogenic substances, since some researchers have discovered that cancer-causing substances can be produced through natural processes as well as through industrial ones, and it has left most ordinary people hopelessly confused. Many simply shrug and say why worry if *everything* can cause cancer. In this case, however, it is important to recognize the crucial differences between natural and synthetic hormone mimics. Many of the man-made hormone mimics pose an even greater hazard than natural compounds because they can persist in the body for years, while plant estrogens might be eliminated within a day.

Regardless of whether natural or man-made, there is reason to be cautious with all hormone-disrupting chemicals. It is true that humans have adapted over millions of years to the presence of hormone mimics in many food plants. But while we may have evolved ways to coexist with such compounds, this does not mean they are harmless. One should never lose sight of the reason plants make them: to sabotage fertility. Our ancestors took advantage of these potent chemicals by using giant fennel, wild carrot, and pomegranate for birth control and abortions. Even naturally occurring hormone mimics can disrupt development of the unborn or young children. Based upon animal studies concerning the developmental effects of plant estrogens, Hughes questions the wisdom of baby formulas containing soy, which harbors estrogenic compounds, until more comprehensive studies are completed.

If some scientists are seeking to identify hormone-disrupting chemicals, others are exploring the hazards they might pose.

In the darkened laboratory at the U.S. Environmental Protection Agency's Health Effects Research Laboratory in Research Triangle Park, North Carolina, slides showing rat genitals, scrambled gonads, and all manner of sexual confusion flash onto the screen. The man at the control button is Earl Gray, a reproductive toxicologist, who makes his living by studying how chemicals disrupt sexual development. He is describing all the ways that synthetic chemicals wreak havoc with hormones and showing the consequences to those exposed in the womb. His fascinating and disturbing slide show brings to mind the title of the Paul Simon song but with a slight revision. There must be fifty ways to lose your fertility—or more.

He stops on a picture of a rat abdomen, that of a male rat who looks like a female. Gray points out the pink nipples protruding through the rat's white fur. Male rats in this breed aren't supposed to have any nipples at all. This male's sexual development went awry because its mother was exposed during pregnancy to vinclozolin, a synthetic chemical that is widely used to kill fungus on fruit. Vinclozolin was frequently detected in the foods children commonly eat in the United States.

Vinclozolin disrupts development as dramatically as DES or other estrogen mimics but causes its havoc in a different way. First of all, it targets the androgen receptor, which is tuned to the male hormone testosterone, rather than the estrogen receptor. Like the mimics, this chemical occupies the receptor, but unlike them, it does not turn on the biological response normally triggered by testosterone. Instead, vinclozolin simply blocks the receptor and doesn't let the testosterone messages through. This is like jamming the line on a cellular phone so it is always busy and the intended messages are blocked. Without these testosterone signals, male development gets derailed and boys don't become boys. Instead, they become stranded in an ambiguous state, where they cannot function as either males or females. In scientific terms, these are "intersex" individuals or hermaphrodites—a term that comes from the Greek deity Hermaphrodite whom classical sculptors portrayed as a figure with male genitals and female breasts.

Gray and his colleague William Kelce have also recently discovered that DDE, a ubiquitous chemical and the DDT breakdown

product found most often in the human body, acts as an androgen
blocker. Like vinclozolin, it binds to and blocks the androgen receptor, so the body's own signals do not get through. Gray believes there
are more anti-androgens to be discovered and more out in the environment than anyone has suspected.

Gray is trying to drive home a point that is often overlooked:
estrogen mimics are only one manner of hormone disruption, only
one of the hazards to sexual development and fertility. All too frequently, the threat from hormone-disrupting chemicals is seen solely
as a problem of estrogen mimics. This is perhaps understandable.
Because of the DES experience, scientists have studied man-made
chemicals that can bind to the estrogen receptor for more than two
decades, describing everything from the action on the cellular level
to lifelong impacts on humans exposed in the womb. In any discussion of the potential hazard, DES is bound to serve as the principal
reference point. There is certainty about how it works.

The obliging nature of the estrogen receptor is another reason
estrogen mimics receive a good deal of attention. Scientists never
did find a simple structural characteristic common to all the foreign
chemicals that the estrogen receptor consorts with, though they
speak vaguely about the fact that these molecules are flat or "pla-
nar." The fact remains, however, that the estrogen receptor binds to
many chemicals with strikingly different structures.

The politics of the breast cancer issue have also helped push estrogen to center stage. Since estrogen exposure increases the risk of
breast cancer, researchers have been exploring links between breast
cancer rates and estrogenic compounds that accumulate in breast
tissue and other fatty parts of the body. Some grassroots advocacy
groups have seized upon synthetic chemicals as the leading suspect
for what is behind the steady one percent a year increase in breast
cancer rates since World War II.

But there are dangers in focusing so narrowly on estrogen,
warns Linda Birnbaum, the head of the environmental toxicology division at the EPA's Health Effects Research Laboratory. Estrogen is
just one component in the complicated, integrated endocrine system, and, she says, synthetic chemicals target other parts of the system more commonly than they disrupt processes involving estrogen.

The adrenal glands, which produce stress hormones, get hit more than any other organ by man-made compounds, followed by the thyroid gland. Insults in any part of one system tend to quickly ripple through other systems of the body as well. So while breast cancer could be linked to estrogenic pesticides, it could also be linked to other kinds of hormone disruption. Birnbaum notes, for example, that depressed thyroid levels have been linked to breast cancer just as increased estrogen exposure has.

However important, estrogen and the receptor mechanism are far from the whole story on endocrine disruption. Man-made chemicals scramble all sorts of hormone messages, and they can disrupt this communication system without ever binding with a receptor. If cellular phone messages aren't getting through, the problem isn't necessarily with your phone. There may be trouble somewhere else in the system, such as in the satellite that relays the message from continent to continent or the transmitter that sends the message into space. The same holds true with the endocrine system.

"If we're thinking only in terms of estrogenicity, we're missing the boat," Earl Gray warns.

For example, another large class of fungicides, members of the pyrimidine carbinol family, inhibit the body's ability to produce steroid hormones in the first place, so vital messages are never sent. Curiously, they interfere with hormone production for exactly the same reason they prevent fungi from growing, by inhibiting the synthesis of fatty compounds called sterols. The fungus needs these fatty substances to form cell membranes, and without them, its growth screeches to a halt. Humans and other mammals form steroid hormones from a much talked about member of this same chemical family, cholesterol.

And even within the group of compounds known to disrupt estrogen levels, other mechanisms can be at work. Although DDT is regarded as a classic estrogen mimic that elevates hormone levels, this is only one of its effects in the body. According to Gray, DDE, the form of DDT that persists the longest in the body fat of humans and animals, has the opposite effect. It depletes hormones by accelerating their breakdown and elimination, leaving the body short not just of estrogen but of testosterone and the other steroid hormones as well. This can lead to abnormally low hormone levels. Since a developing fetus is extremely sensitive to hormone levels, too little can be as devastating as too much.

On the other hand, foreign chemicals that do not act as hormone mimics or blockers may boost the body's hormone levels by interfering with the physiological processes that break down hormones so they can be excreted. Some chemicals deactivate the enzymes involved in this process, according to Michael Baker, an enzyme specialist at the University of California at San Diego. If some chemical interfered with the enzyme that helps break down estrogen, for example, it would cause more estrogen to be available to the receptors and indirectly create an estrogenic effect without binding itself to the receptor. Based on the body's response, one might mistake such a chemical for a hormone mimic.

In Earl Gray's view, animal studies of hormone-disrupting chemicals have clear and immediate relevance to humans. In the broader environmental debate, some have challenged the predictive value of rat studies to assess possible cancer risks posed by synthetic chemicals to humans on the grounds that animals and humans sometimes react differently to a chemical. The use of animals to study hormone-disrupting chemicals is, however, fraught with less uncertainty, Gray explains, because scientists understand far more about the role of hormones in development than they do about the biological events that give rise to cancer. Moreover, the evidence shows that humans and animals respond in generally the same way to hormone-disrupting chemicals. The available human data and the effects seen in lab animals show "a perfect correlation." Earl Gray spells out the bottom line with intensity and directness.

"We know a lot about the process. We know it can be altered by chemicals. It is important to take the effects you see in animal studies seriously."

TO THE ENDS OF THE EARTH

AFTER MORE THAN THREE MONTHS OF DARKNESS AND TEASING TWI-light, the sun finally shoulders its way above the horizon, signaling the approach of spring on Kongsøya Island, high above the Arctic Circle in Norway's Svalbard archipelago. As the days quickly lengthen, ringed seals begin to venture out of the water and dig nursery lairs in the snowdrifts on the ice near shore. One by one, the polar bears stir in their dens beneath the deeply drifted snow. Only pregnant fe-males hole up for the winter, while females that are not pregnant and males spend the months of darkness wandering great distances over the shifting pack ice that surrounds this part of Svalbard for most of the year.

Kongsøya, a rocky, treeless island lying east of Greenland at sev-enty-nine degrees north latitude, is something of a maternity ward for the great white bear. Pregnant females often seek out the island's Bogen Valley, digging their nursery dens on its south-facing slopes as winter sets in. There, in hibernation during the long winter night, they give birth to one-pound cubs and nurse them for several months before reemerging with twenty-pound youngsters in spring.

Twelve hundred miles to the south in Oslo, Øystein Wiig, a polar bear researcher, keeps tabs on more than a dozen of Svalbard's pregnant bears from the warmth of his office at the Zoological Museum. Much about these bears has remained a mystery because of the cold, the darkness, and the remoteness of the high Arctic. But now, with the help of modern technology such as helicopters and satellites, scientists like Wiig can probe their previously hidden lives. On field trips to Svalbard, he has fitted the female bears with radio collars that beam signals to passing satellites that then transmit the information to Wiig. As the bears leave their dens in March, Wiig and his assistants head for Svalbard to track the new mothers and learn more about the two thousand bears who roam the archipelago.

Based on what researchers had already learned, Wiig expected at least twelve of the hibernating females to produce offspring, most likely twins. But to his surprise, only five of the pregnant bears emerged from their dens with young in 1992.

A single bad year should be no cause for alarm, for a female's success in producing offspring can depend on many things, including ice conditions, weather, the supply of seals for food, and the density of the bear populations. The age at which female polar bears first reproduced in Svalbard had increased by a year over the past decade or so, reflecting, Wiig speculates, that the bear population is reaching its limit given the available food supply. But the failed pregnancies were nevertheless worrisome, given what Norwegian researchers are finding in the fat of the polar bears. Though Svalbard is remote and seems pristine, the bears there are highly contaminated with industrial chemicals, including PCBs, the pesticide DDT, and several other persistent man-made compounds that are known to disrupt reproduction in wildlife. Wiig and his colleagues collect the fat from live bears by shooting them with a dart containing a tranquilizer. Once a bear becomes immobilized, Wiig might fit it with a radio collar and bore into its blubber using a device similar to an apple corer.

Some Svalbard bears carry as much as ninety parts per million of PCBs in their fat—an infinitesimal amount by normal measures, but biologically a potent dose. Researchers studying declining seal populations have found that seventy parts per million of PCBs is enough to cause serious problems for females, including suppressed immune systems and deformities of the uterus and of the fallopian tubes that transport the eggs from the ovaries. But those seals were living in Wadden Zee, off the Netherlands, where industrial waste has poured in for decades from Europe via the Rhine-Meuse estuary. Svalbard, on the other hand, lies at the end of the Earth, hundreds of miles from cities, chemical factories, farm fields, and dump sites.

Where was this stuff coming from, and how was it getting into bears roaming the Arctic wilderness?

The story of PCBs and how they have spread throughout the planet and into the body fat of almost every living creature is one of the most fascinating and instructive chapters in the history of the era of synthetic chemicals. Of the fifty-one synthetic chemicals that have now been identified as hormone disruptors, at least half, including PCBs, are "persistent" products in that they resist natural processes of decay that render them harmless. These long-lived chemicals will be a legacy and a continuing hazard to the unborn for years, decades, or in the case of some PCBs, several centuries.

Introduced in 1929, PCBs became the first big commercial success for a new elite of chemists who would eventually synthesize tens of thousands of novel chemicals that exist nowhere in nature. The engineers created PCBs by adding chlorine atoms to a molecule with two joined hexagonal benzene rings known as a biphenyl. The result of their tinkering was a family of 209 chemicals known collectively as polychlorinated biphenyls, or PCBs, which soon proved to be immensely useful compounds.

In early assessments, PCBs seemed to have many virtues and no obvious faults. They are nonflammable and extremely stable. Toxicity tests at the time did not identify any hazardous effects. Confident of their safety as well as their utility, the Swann Chemical Company, which would soon become a part of Monsanto Chemical Company in 1935, quickly moved them into production and onto the market.

With the issuance of federal regulations requiring the use of nonflammable cooling compounds in transformers used inside buildings, PCBs quickly found a steady major market in the electrical industry. Other industries put PCBs to use as lubricants, hydraulic fluids, cutting oils, and liquid seals. In time, these chemicals also found their way into a host of consumer products and thus into the home. They made wood and plastics nonflammable. They preserved and protected rubber. They made stucco weatherproof. They became ingredients in paints, varnishes, inks, and pesticides. In retrospect, it is clear that the very characteristics that made them a runaway commercial success also made them one of our most serious environmental pollutants.

Although some evidence of toxic effects in workers began emerging as early as 1936, indicating that PCBs were not as safe as previously believed, PCBs were on the market for thirty-six years before serious questions surfaced publicly about this wonder chemical. In the meantime, manufacturers kept coming up with new uses. From 1957 through 1971, paper companies put PCBs in their carbonless copy paper, enabling typists, in an era before widespread use of the copying machine, to make duplicates of documents without carbon paper.

The person to first recognize that PCBs had become a pervasive contaminant was the Danish-born chemist Sören Jensen. In 1964 Jensen, who worked at the Institute for Analytical Chemistry at the University of Stockholm, kept encountering mysterious chemical compounds as he tried to measure DDT levels in human blood. Whatever it was, Jensen found it wherever he looked—in wildlife specimens collected three decades earlier, in the Swedish environment, in the surrounding seas, in hair samples from his wife and infant daughter. The presence of the mystery contaminant in wildlife samples taken in 1935 indicated it could not be a chlorine-based pesticide, which came into broad use only after World War II. It took Jensen more than two years of investigation to identify the synthetic pollutant as PCBs. A report of Jensen's findings first appeared in the British journal *New Scientist* in 1966.

As other scientists began to look for PCBs, they, too, found them *everywhere*—in soil, air, water; in the mud of lakes, rivers, and estuaries; in the ocean; in fish, birds, and other animals. Chemists had long puzzled about the elusive peaks that showed up repeatedly on their gas chromatograph charts when they were analyzing samples taken from the environment. The peaks, which looked similar to those made by DDT, registered the presence of some chemical, but until Jensen compared the peaks using a chemical sample provided by a German manufacturer, they did not know what the contaminant was. Finally, they had the answer—PCBs.

Ten years later, in 1976, the United States banned the manufacture of PCBs, and other industrial countries eventually followed. In half a century of production, however, the synthetic chemical industry worldwide (excluding the USSR) had produced an estimated 3.4 billion pounds of PCBs, and much of it was already loose in the environment and beyond recall. Moreover, the ban did not address existing PCBs, allowing their use to continue in closed applications—such as transistors, electric ballasts, and small appliances—even today.

There is no way to discover *exactly* how the PCBs in the polar bears made their way to the Svalbard archipelago or where they came from. But research over the past two decades has given scientists a good understanding of how PCBs travel through ecosystems and migrate over long distances. Based on this knowledge, it is possible to imagine the journey of an individual PCB molecule. Though the specific route and events in the journey we are about to describe are hypothetical, the plot is a plausible scenario built from historical accounts and a myriad of scientific studies.

Our imaginary PCB molecule—a chemical known among scientists as PCB-153 because of the arrangement of its chlorine atoms—had already been around for some time before it set off on its global wanderings just after World War II ended. As the market for PCBs began to grow in the 1930s, the Monsanto Chemical Works expanded production at its plant in Anniston, Alabama, where it heated a mixture of the chemical biphenyl, a particular form of chlorine, and iron filings to create PCBs. During the spring of 1947, the workers at the Anniston factory made up a batch of PCBs containing fifty-four percent chlorine. As the chlorine gas bubbled through the heated biphenyl, six atoms of chlorine bonded to one of the biphenyls in the tank, and PCB-153 came into being. After an alkali wash and some distillation to purify the newly formed PCBs, Monsanto often sold the compound—which contained not only PCB-153 but dozens of other members of the large PCB family—under its brand name Aroclor-1254.

Almost half a century later, the PCBs made on that spring day might be found virtually anywhere imaginable: in the sperm of a man tested at a fertility clinic in upstate New York, in the finest caviar, in the fat of a newborn baby in Michigan, in penguins in Antarctica, in the bluefin tuna served at a sushi bar in Tokyo, in the monsoon rains falling on Calcutta, in the milk of a nursing mother in France, in the blubber of a sperm whale cruising the South Pacific, in a wheel of ripe brie cheese, in a handsome striped bass landed off Martha's Vineyard on a summer weekend. Like most persistent synthetic chemicals, PCBs are world travelers.

Our imaginary molecule of PCB-153 that would end up in a polar bear in the high Arctic might have made its first trip by train. A few weeks after the manufacture of this molecule, the freight train carrying a shipment of Aroclor-1254 rumbled over the rails in New York State headed for a plant in western Massachusetts where General Electric manufactured electrical transformers.

These ubiquitous metal cans attached to electrical poles were an essential component in the growing grid that sent electricity from generating systems over high voltage power lines and into homes to power lights, radios, vacuum cleaners, and refrigerators—the wonderful new twentieth-century electric conveniences. The transformers made at GE's Pittsfield, Massachusetts, plant reduced the high-voltage current from the transmission lines into the lower voltage required by lights and appliances.

From General Electric's perspective, the PCBs were an ideal insulating coolant for their capacitors and for transformers used in situations where flammability was a concern. Because PCBs did not catch fire and burn, they offered a safer alternative to the flammable oil used in transformers before this new synthetic product was developed. The company had developed its own custom formula for transformers called Pyranol, containing Aroclor and oils which it blended at the Pittsfield plant.

With the postwar economic boom, the demand for transformers and other electrical equipment seemed insatiable. America was building new houses for returning GIs as fast as it could—houses that needed new appliances and increased electrical service. While it would be very difficult to retrace the precise sequence of events that led to the escape of our PCB molecule into the environment, we can imagine that the next step in its journey took place at Pittsfield, and from interviews with a former plant employee and public records, we can reconstruct what might have taken place on a typical summer day. That summer the production line in Pittsfield was working at full tilt, and the Pyranol in the factory storage tanks did not sit around for long. On a steamy day in June, a worker reached for a hose at his workstation that was connected through underground pipes to storage tanks. After making a final check on the transformer he had been finishing, he opened the valve and filled it to the top with Pyranol. In a few days, our molecule of PCB-153, sealed tightly inside that new transformer, was heading back south by train.

The oil refineries in the west Texas city of Big Spring were also scrambling that summer to keep up with the postwar economic boom, for the explosive growth of suburbs was creating a new class of commuters who needed new cars and gasoline to power them. One of the city's smaller oil and chemical companies, now no longer in business, was moving as quickly as possible to build a new refinery complex, but the project had stalled for several months while the contractors awaited the arrival of back-ordered electrical equipment. The shipment of transformers from GE to the refinery finally arrived in July. Within a week, the distribution transformer containing the molecule of PCB-153 was installed and in service in a building that housed the control room for the new installation.

Not even a month had passed before a fierce August thunderstorm tore through Big Spring, filling the air with exploding thunder and leaping lightning that struck at several places during the short, violent storm, including the power lines supplying the refinery. As the power surge hit the transformer near the control room, it responded with a metallic thump and the building went dark.

The following morning, the refinery's maintenance supervisor lifted the cover of the transformer to inspect the damage. Seeing twisted, crumbled coils, he decided that the unit was beyond repair, so he asked one of his men to empty the unit and send it off to the dump. The maintenance worker complied, hauling the transformer to the parking lot. As he tilted the transformer, its oily contents oozed out onto the red dirt of the parking lot, and PCB-153 slipped into the greasy puddle. The worker reckoned the oil might help keep down the insufferable dust. Since PCBs have an affinity for organic matter, the molecule quickly attached itself to a dust particle.

But with the roaring winds of west Texas, dust never stays put long. Four months later, a winter storm roared through and swept the molecule aloft. Stampeding curtains of dust drove toward the town of Tarzan, where they beat against barns and houses as the winds howled. The dust particle with PCB-153 rode the whirlwind, bouncing with the turbulence like a cowboy on a bronco. The wild ride ended when the dust particle sifted through the fine cracks around a doorsill and settled in a drift on the kitchen floor.

When the windstorm passed, the woman of the house surveyed her kitchen with a sigh. The fine red dust coated the windowsills and lay two inches deep before the door. With a weary efficiency, she took up her corn-straw broom and whisked the dust particle with our itinerant molecule into a dustpan. As it fell into the wastebasket, the dust particle sifted down into a crumpled, grease-stained newspaper page that the housewife had used to drain her bacon that morning.

By the end of the week, PCB-153 was buried under trash in a local dump, an informal affair in a ravine with a parched creek bed. Despite the rivulets that flowed down through the growing mountain of trash during summer thunderstorms, the molecule stayed put for more than two years, for unlike many chemicals, PCBs don't dissolve readily in water.

The late winter of 1948 brought a spell of heavy rains to west Texas. After intermittent downpours, the creek surged to life in the beginning of March and roared toward the trash that tumbled down the side of the ravine. The roiling waters took a bite out of one edge of the trash mound, exposing a cross section of the town's recent history and sweeping the greasy newspaper and the molecule from the transformer spill downstream. The floodwaters subsided the following morning, leaving the soggy newspaper sheet stranded on a sandbar five miles away. PCB-153 was clinging to a greasy blotch on the page, shielded from the light but exposed to warm spring air.

As the sun climbed higher and winter turned to spring, the lump of paper dried and slowly warmed. With the sun beating down on the paper in early April, PCBs suddenly began disengaging from the dust particle moving upward, floating into the air as a vapor. The PCB-153 was suddenly free. The journey that would end in the rump fat of a Norwegian polar bear had begun.

The molecule caught a warm gentle breeze from the southwest, wafting north and east over the shrub-covered expanse of east Texas toward the fragrant pine forests of Arkansas. As the breeze stiffened, it sailed on unimpeded into Missouri. A rising current of spring air pushed it higher into the atmosphere, and the molecule soared upward, higher and higher on the thermal. When the air mass collided with a cold front moving down from the north, the journey ended abruptly. The clouds released their moisture in a hard, cold rain, and PCB-153 washed back to earth and landed on a bluff overlooking the Mississippi River north of St. Louis.

During three weeks of unusually cool and cloudy weather, the molecule clung to a rotting leaf in a hollow on a rocky outcropping, but as soon as the sun reemerged and the temperature climbed, the molecule floated off again. It lingered over St. Louis for several days and sloshed about in a stagnant air mass. Then, as a Bermuda high developed off the southern Atlantic coast, a torrent of air rushed through from the south and swept the PCB molecule northward over the radiant green cornfields of southern Illinois toward the Great Lakes.

The air flow generated by the Bermuda high pushed north with the speed of a freight train. The molecule tumbled onward in a great white bank of cumulus clouds. But as the warm winds rushed through Chicago and out over Lake Michigan, they met a wall of cooler air because the Great Lakes, like any large body of water, warmed more slowly in spring than the surrounding land. In the night chill, the PCB-153 suddenly condensed back into a liquid state for the first time since it had left Missouri.

The breeze died just before midnight and the molecule settled on the dark water near the lakeshore city of Racine, Wisconsin. Like all PCBs, the molecule had a predilection for surfaces, so it lingered about on the boundary between the air and water, bumping now and again into other wandering members of its extended family. The molecule found it hard, however, to remain unattached for very long. Its strong attraction to organic matter drew it to a patch of algae, rootless plants that floated like a gauzy green veil near the water's surface. When the opportunity presented itself, the molecule grabbed on to one of the tiny plants and, clinging to its waxy surface, washed back and forth along the shoreline near the mouth of the Root River, moving with the vagaries of the wind and waves.

Plant-eaters, such as the water flea, nibbled around the edges of the green veil, and some of the plants ended up as their salad course, but the tiny plant PCB-153 was riding managed to escape and live its full life span—which lasted three weeks. The alga began to yellow and grow tattered around the edges. The dead plant grew water-logged and sank, carrying PCB-153 with it.

The dead alga settled on the bottom and was quickly covered by soil washing into the lake from a city dump at the water's edge. Accumulating sediment buried the molecule ever deeper in the lake muds, and with each passing year, its chances of getting back into circulation seemed to grow slimmer. PCB-153 might be impervious to the attack of the bacteria that broke down most chemicals, but it could be entombed.

Persistence is viewed as a virtue in people. In chemicals, it is the mark of a troublemaker. The synthetic chemical industry helped bring convenience and comfort to American homes, but at the same time, it unleashed dozens of chemicals, including PCBs, that became notorious for combining the devilish properties of extreme stability, volatility, and a particular affinity for fat.

Besides PCBs, this lot includes the pesticides DDT, chlordane, lindane, aldrin, dieldrin, endrin, toxaphene, heptachlor, and the ubiquitous contaminant dioxin, which is produced in many chemical processes and during the burning of fossil fuels and trash. They ride through the food web on particles of fat or vanish into vapors that gallop on the winds to distant lands. In *Silent Spring*, Rachel Carson put the persistent pesticides at the top of her most-wanted list. It didn't occur to her to include compounds such as PCBs that may not be particularly poisonous (in the usual sense of causing immediate death or cancer) but are persistent—a fact scientists did not recognize until 1966, four years after *Silent Spring* was published.

The members of the PCB family that contain fewer chlorine atoms do have a few enemies, including two bacteria from the *Achromobacter* genus. But chlorine heavies like PCB-153 are impervious to almost everything save ultraviolet B radiation from the sun. Because of the way PCBs move through the environment, hidden in soil, sediments, animal tissue, and other places, these deadly rays only rarely encounter them.

With the prosperity that followed the end of World War II, the lakefront in Racine began to change. For half a century, the gas plant with its huge black storage cylinders and mountains of coal had loomed over central Racine and the lakefront. With the arrival in Wisconsin of a long-planned natural gas pipeline from the southwest in 1949, the old facility that made cooking and heating gas from coal became history.

A few years later, Racine took its first step toward reclaiming its lakefront for pleasure—a vision that it would not fully realize until the 1980s. Nevertheless, the city's first waterfront park was, indeed, a significant beginning, since the plans called for converting the now-abandoned city dump into a twenty-seven-acre recreation area with trees, grassy playing fields, a boat ramp, and a scenic shore drive. At the request of American Legion Post 76, the city agreed to christen it Pershing Park after a World War I hero, General John J. Pershing.

During the construction of Pershing Park, which commenced in 1954, the work crews began to fashion the new shore where the scenic drive would run. Dump trucks laden with huge chunks of rock moved back and forth in a grumbling caravan to the water's edge. Load after load of stones tumbled into water. On a spring day in 1956, a monstrous rock came hurtling down into the stretch of sediment where the PCB 153 lay buried. As the shock rippled through the buried mud, the molecule sprang free in a burst of hydrogen sulfide gas that filled the water with bubbles. The molecule rode one of the tiny glistening spheres upward toward the light and air.

Within hours, the PCB-153 was lodged in the fat of a water flea that had gobbled it up while grazing along the surface. This was a fat-loving molecule's dream, a ticket to ride, and it would take PCB-153 to the top of the food chain.

The water flea acted like a filter, sifting tiny plants and the PCBs clinging to them out of the water while it fed, so as the days passed more and more PCBs accumulated in the body fat of the tiny animal. Less persistent contaminants do not build up in this way because animals can break them down into water-soluble substances and excrete them. Many PCBs, on the other hand, resist breakdown, and once ingested, they are drawn by their chemical structure to the animal's fat, where they remain indefinitely. Over its short ten-day life, the flea's PCB concentrations grew to four hundred times the levels in the water. When it was finally eaten by a small shrimp called a mysid, the water flea passed on this legacy of fat-loving persistent chemicals to its predator, and the PCB-153 moved a rung higher on the Lake Michigan food web.

In its life, the mysid would eat hundreds of water fleas and inherit a bundle of persistent chemicals with each bite. The PCBs riding in its fat, like PCB-153, found themselves in growing company that included not only their own chemical family but other persistent compounds as well, such as DDT and toxaphene, a pesticide used heavily then in cotton fields in the south. For a time, agricultural specialists thought toxaphene was an improvement over DDT because it disappeared quickly from the fields where it was sprayed. It was some time before anybody realized that it did not disappear at all; it vaporized and moved on. Much of it came sailing on the winds to the Great Lakes.

The mysid eventually became a meal for a smelt—a small, tasty fish that darts about in offshore waters in flashing silver schools. As the smelt gorged on mysids and other smaller critters, the persistent chemical concentrations multiplied seventeen more times.

Smelt were a particular favorite with the families that crowded into local eateries in Racine for the Friday-night fish fry, and over time, their body fat would also bear witness to the nights when they had popped down dozens of the succulent little fish with a side order of potato pancakes or dined grandly on lake trout or coho salmon.

The smelt containing our PCB-153 molecule cruised Lake Michigan for two years before it was ambushed by a lake trout. Now the molecule moved to the trout and rested in its fat for another five years until an angler hooked the trophy-size fish on the last day of vacation at a family cottage in Door County, Wisconsin.

The following morning, the trout, packed on ice in a cooler in the back of a station wagon, was heading eastward on the interstate toward upstate New York. The molecule was moving into new territory on this imaginary journey. The fisherman could hardly wait to get home and show off the catch of a lifetime to fishing buddies. His mouth watered at the thought of a truly memorable fish dinner with his family.

Three days later, however, the fish ended up in the family's trash barrel rather than on a platter at the dinner table. At the height of an August heat wave, the station wagon had broken down, leaving the family stranded at a gas station in rural Michigan without transportation and without ice for the cooler. When the family reached home and opened the cooler, the fish smelled like old cat food.

A blizzard of gulls swirled around the trash collection truck when it arrived at a landfill outside Rochester. As the rank fish carrying the molecule tumbled onto the growing trash mountain, the gulls dove at it like shoppers at a half-price sale, squawking and jostling each other to grab a bite. In a matter of minutes, they had picked the carcass clean.

PCB-153 wound up in the fat of a female gull, which had spent more than a dozen years feeding on the fish in Lake Ontario, so the molecule simply added to her already substantial store of contaminants. In the Great Lakes food chain, the herring gulls occupy a spot just below the bald eagles, which sometimes nab a herring gull or two. By the time PCBs have moved this high on the food chain, the concentrations have multiplied to 25 million times the levels found in the water.

The following spring, the female herring gull headed for Scotch Bonnet Island, roughly one hundred miles east of Toronto on the Canadian shore of Lake Ontario. The gull and her mate quickly set their stake on a good patch of sand in the middle of the gull colony, a location considered preferable to the edges, where the chicks might be more vulnerable to predators. The pair courted and mated. Afterward, the female scraped a hollow in the sand and laid two large lightly speckled eggs that she dutifully set about incubating.

A tiny beak broke through one shell six weeks later, but the chick could only muster feeble pecks and it died, seemingly from exhaustion. The other egg showed no signs of life at all, but the pair stayed on the nest for another week. The mother finally abandoned the nest without fledging a single offspring.

PCB-153 and its relatives had passed from the mother gull into the yolk of the lifeless egg and had contributed to its death, along with DDT, dioxin, and other contaminants. A skunk carted off the rotting egg five days later but then thought better of eating it and dropped it on a rock near the shore, where it smashed. Some of the yolk spattered into the water, and PCB-153 was off on another trip up the food chain, this time via a crayfish—a small bottom-feeding scavenger that vacuumed up the bits of fatty yolk sloshing in the shallows near shore.

Before long, the crayfish that had dined on the egg yolk became dinner for one of the American eels that hunted at night in the weedy shallows. The eel is something of a contrarian when it comes to spawning. Many species, such as salmon and herring, spend their maturity at sea and then return upriver to their birthplace to spawn. American eels, on the other hand, frequent freshwater rivers and lakes most of their lives before finally making a long pilgrimage out to the Sargasso Sea—an area in the Atlantic Ocean between the West Indies and the Azores—to spawn before dying. Curiously, the eels migrating from the Great Lakes and other northern waters are all female, while those migrating from southern rivers tend to be males.

With the approach of summer, the oldest eels in Lake Ontario, including the sixteen-year-old animal carrying the PCB-153 molecule, began to undergo changes that signal sexual maturity and preparation for the three-thousand-mile journey to the spawning ground. Their gray green backs began to darken toward silvery black, their yellow bellies whitened, and their eyes enlarged and changed to allow better vision in deeper ocean waters. Restless with the migratory urge, groups of silver eels moved toward the entrance to the St. Lawrence River waiting apparently for some sign that the time was right. Then on a stormy night when rain poured in dense silver sheets out of a black sky, a slithering multitude suddenly departed down the great river toward the North Atlantic. The eel carrying PCB-153 swam onward for more than six months with a mystifying urgency before finally reaching the floating rafts of sargasso seaweed that give this region of warm, salty waters its name. There beneath the clear tropical waters east of the Bahamas and south of Bermuda, a roiling congregation of eels, gathered from the Gulf Coast to Newfoundland, spawned and then expired from exhaustion—their long journey ended, their compelling mission accomplished.

The eel's flesh disintegrated quickly in the warm tropical waters, and PCB-153 sloughed off in a shred of fat that floated up to the surface of the Sargasso Sea under the intense tropical sun. In the heat, the molecule suddenly vaporized once more and, carried on prevailing winds, began hopscotching north. At any cold spot it encountered, the molecule condensed and settled on any available surface, only to be off again as soon as the summer sun warmed the surface. Alternating between liquid and gas, it rode the winds farther and farther north. The waters grew colder, making it increasingly difficult for the molecule to become airborne. Instead it hitchhiked on one of the small floating plants at the bottom of the North Atlantic food web, sweeping into the Gulf Stream and from then on north and east toward Iceland.

Two hundred miles east of Iceland, a small shrimplike creature called a copepod finally nabbed the plant and PCB-153 as it filtered a meal out of the rich waters of the North Atlantic. Five days later, a cloud of copepods was swept into a swift current that carried it quickly north and east like a giant conveyer belt toward the edge of the solid pack ice in the Greenland Sea, where a large school of Arctic cod had gathered to feast on the incoming bounty.

The gray green water boiled with the feeding cod, one of the most abundant species in high Arctic waters. As one of the small fish digested its stomachful of copepods, PCB-153 migrated to the fatty tissue near its tail, which already had a considerable store of persistent chemicals. The Arctic food web, which includes the cod, is quite simple, but it includes many long-lived animals that accumulate significant amounts of contamination over a lifetime. For this reason, the Arctic food web concentrates and magnifies persistent chemicals to an even greater degree than that of the Great Lakes. Though far from a top predator, this cod carried PCBs at 48 million times the concentration found in the surrounding waters. Even so, the cod are still less contaminated than Great Lakes salmon because the ocean waters they inhabit are far cleaner.

Arctic cod spend the greater part of their lives feeding beneath the solid vault of ice that closes over high Arctic waters for most of the year. That season, the cod carrying PCB-153 followed the shifting food supply, and on the trail of a particularly abundant crop of copepods, it gradually swam toward the eastern part of the Greenland Sea. During the icebound period, ringed seals depend exclusively upon the schools of cod that wander beneath the ice.

It was only a matter of time before the cod carrying PCB-153 became a meal for a hungry adolescent seal that shot through the water, propelled by its powerful hind flippers. Like many seals searching for food, the youngster had wandered along a fracture in the sea ice west of the Svalbard Islands. The hunting had been good that winter, and the seal had added significantly to its ample blubber, which, despite its short life, contained not only PCB-153 but a high concentration of chlordane, DDT, toxaphene, and other persistent chemicals that were finding their way to the Arctic from all over the world. A seal eats hundreds of fish, ingesting and storing all the PCBs that had accumulated in them. For this reason, the PCB levels in the seals are eight times greater than in the cod, or 384 million times the concentrations in the ocean water.

Once the sea closes over with ice, the seals breathe through holes that they keep open by punching through at regular intervals with their noses. The great white bears can sniff out these holes from a remarkable distance, and they often hunt by waiting in ambush.

The young seal had just surfaced to breathe when a bear that had been waiting on its stomach near the breathing hole lunged out and, in a single continous movement, flipped the 150-pound animal out of the water and onto the ice. The ringed seal died instantly in the attack by the five-year-old female bear, who had learned the hunting technique from her mother before going off on her own two and a half years earlier.

In thirty minutes, she had consumed the best parts of the seal— its skin and succulent blubber—and acquired PCB-153 along with a considerable synthetic chemical legacy. The bear was quickly gaining weight because of the good hunting, so as she laid on more fat the molecule moved into her well-insulated rump.

As spring advanced, the young female gradually made her way toward the land-fast ice around Svalbard, which provides the best hunting of the year. The seals are particularly vulnerable when they haul out onto the ice to have their pups, so for bears the living is easy. Many gather in this polar bear fat city, feasting for days and consuming huge quantities of seal blubber. Growing visibly rotund, some animals triple their body weight.

In late April, the young female mated for the first time, with a large male who had also come to Svalbard to feed on the seals. But as is the case with all bears, the fertilized eggs did not begin developing immediately. Instead the female carried them around in her body until the following November, when she hollowed out a den in the growing banks of snow on Kongsøya Island. As she settled in for the winter, the fertilized eggs implanted in her womb and began to grow. In the dead of winter, two tiny pink cubs, weighing only a pound and a half each, slipped into the world unnoticed by their sleeping mother. Crawling across her creamy expanse, they found their way to her nipples and began nursing on her rich, fatty milk.

Throughout the winter, the mother and cubs all lived on the ample layers of fat she had laid down the previous year. As the fat melted away, PCB-153 was on the move again, this time into the breast milk of the female bear. The cubs were nursing greedily and growing rapidly. As one of the female cubs tugged at the nipple, the molecule shot into her mouth with a blast of warm, thick milk

No one yet knows how persistent chemicals like PCBs harm polar bears or how much it takes to cause damage. But given experience with other wildlife species, it seems certain that PCB-153 and other persistent hormone-disrupting chemicals pose a greater hazard to the developing cubs than to the mother who ingested the chemicals from the seal blubber.

The polar bear twins topped twenty pounds by the time the new mother emerged from her den into the honey light of the Arctic spring. They would continue to nurse for more than two years and grow to roughly four hundred pounds each on the rich diet of polar bear milk. With each meal, they would take in more of the persistent

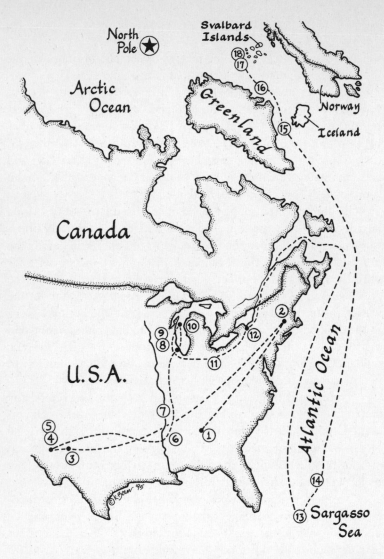

chemicals that had traveled thousands of miles to the remote Arctic. The concentrations of PCBs had multiplied 3 billion times as they moved up the Arctic food chain to the polar bear, the top predator and largest land carnivore.

A decade later, one of the twins may have been among the pregnant females who emerged from a den in Svalbard without any cubs of her own.

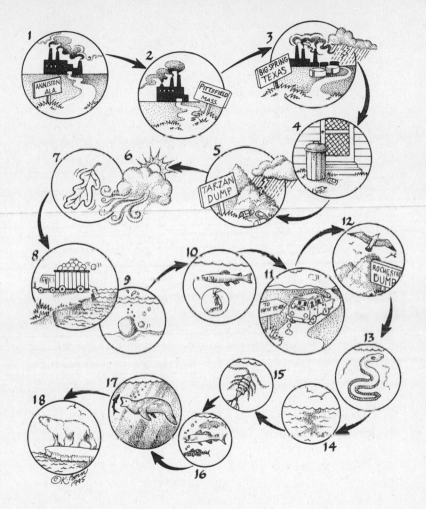

Chemicals manufactured on one continent can travel thousands of miles away. This path traces the journey of a PCB molecule from its point of origin in a factory in Alabama to a refinery in Texas and up the food web in the Great Lakes and North Atlantic regions. The concentration of persistent chemicals can be magnified millions of times as they travel to the ends of the earth.

* * *

Like polar bears, humans share the hazards of feeding at the top of the food web. The persistent synthetic chemicals that have invaded the great bear's world pervade ours as well.

Humans also carry PCBs and other persistent chemicals in their body fat, and they pass this chemical legacy on to their babies. Virtually anyone willing to put up the $2,000 for the tests will find at least 250 chemical contaminants in his or her body fat, regardless of whether he or she lives in Gary, Indiana, or on a remote island in the South Pacific. You cannot escape them. Ironically, some of those living farthest from industrial centers and sources of pollution have suffered the greatest contamination: these chemicals travel long distances and build up along the way to high concentrations, especially in the Arctic, which is becoming a final resting ground. These synthetic chemicals move everywhere, even through the placental barrier and into the womb, exposing the unborn during the most vulnerable stages of development. When a new mother breast-feeds her baby, she is giving it more than love and nourishment: she is passing on high doses of persistent chemicals as well.

It has been three decades since health researchers discovered that DDT, PCBs, and other persistent chemicals were accumulating in human body fat and breast milk, as well as in every other part of the environment. The measurements have been the easy part. Since then, concerned scientists have been trying to understand their meaning. If we all carry around an alphabet soup of novel chemicals in our body fat, how is it affecting us? How is it affecting our children?

While researchers do not have all the answers to these questions, they are convinced that humans carry high enough levels of synthetic chemicals to endanger their children. Without knowing exactly how all these chemicals act, separately or together, the researchers have linked them not only to damage in wildlife offspring but in humans as well. We explore these links in later chapters.

While prenatal exposure seems to pose the greatest hazard, health specialists also worry about the chemicals passed on in breast milk because some sensitive developmental processes continue in the weeks immediately after birth. During breast feeding, human infants are exposed to higher concentrations of these chemicals than at any subsequent time in their lives. In just six months of breast feeding, a baby in the United States and Europe gets the maximum recommended lifetime dose of dioxin, which rides through the food web like PCBs and DDT. The same breast feeding baby gets five times the allowable daily level of PCBs set by international health standards for a 150-pound adult.

The contamination of breast milk has been particularly severe among indigenous people in the high Arctic, where many people still eat the wild food the land and sea provide. There, researchers have found that babies take in seven times more PCBs than the typical infant in southern Canada or the United States. The PCBs and other chemicals that contaminate the infants have almost all arrived by wind and water currents.

Canadian health officials have noted that many children in Inuit villages are plagued by chronic ear infections. Recent studies have found abnormalities in the immune systems of these children, including the discovery that their bodies do not produce the necessary antibodies when they are vaccinated for smallpox, measles, polio, and other diseases. The failure of vaccinations could make these children much more vulnerable to disease.

The Inuktitut language spoken on Broughton Island in the Canadian Arctic contains no word for contamination. This has made it all the harder for the Inuit people living there to grasp the news brought by Canadian health officials that persistent synthetic chemicals are polluting the high Arctic and the food they eat. Perhaps, some villagers suggested, the government officials were telling them the animals had something called PCBs to scare them, to keep them from killing any more whales or polar bears. Perhaps they were in cahoots with the animal rights crowd.

Broughton Island, which has a village of 450 people, lies off Baffin Island, west of Greenland, more than sixteen hundred miles from the smokestacks of southern Ontario, and twenty-four hundred miles from the industrial centers in Europe, but that distant world has cast its long shadow over the villagers' lives, filling them with uncertainty and fear. It threatens their culture, which has endured for thousands of years.

As their ancestors have before them, the men of Broughton Island fish and hunt to put food on the table. While they may give
chase these days by snowmobile and power boat instead of dogsled
and kayak, they still pursue seals, polar bears, caribou, and narwhals—
small whales with a spiral tusk on their head like the legendary uni-
corn. The island does have a store that sells imported food, but the
diet of most islanders still consists largely of wild fish and game.

As the Arctic has become the resting place for volatile persistent chemicals, the contamination has passed up the food web
to humans. Canadian health studies have shown that the people
on Broughton Island have the highest levels of PCBs found in
any human population except those contaminated in industrial
accidents.

The provincial health officials have told the villagers about the
contamination found in their bodies, but they have not been able to
tell them what these high PCB levels mean to their health or to the
health of their children. In the meantime, they have recommended
that the villagers continue to eat the traditional Inuit diet, which is
otherwise far more nutritious than the food imported by bush plane
and sold at a high price in the village store. In any event, with milk
going for $4 a bottle and small turkeys for $40 each, most villagers
have little choice.

Whatever the health effects, the report of high PCB levels,
which was widely covered by the Canadian press, has caused economic, social, and psychological turmoil for the Broughton Islanders.
Apparently unaware that they are probably carrying high PCB levels
as well, other Baffin Island Inuit communities have begun shunning
the villagers as the "PCB people" and discouraging marriages to
them. A fish dealer in the south, who used to buy and sell Arctic char
caught by the men of Broughton Island as a gourmet specialty, canceled his contract, thus cutting off one of the major sources of the islanders' cash income.

The news that their breast milk contains chemicals has left
some of the women frightened and desperate. One mother decided
to stop nursing in an effort to protect her new baby. After several
weeks of being bottle fed a mixture of water and Coffee-mate, the
baby was hospitalized.

The Broughton Island people are not a unique case, only the most extreme example discovered thus far of human contamination with persistent chemicals. No matter where we live, we share their fate to some degree. Many chemicals that threaten the next generation have found their way into our bodies. There is no safe, uncontaminated place.

7

A SINGLE HIT

Out in the wild, Theo Colborn quickly learned, persistent chemicals were showing up in the most unexpected places. By early 1990, she knew that the problem extended far beyond the Great Lakes. Her file cabinets contained dozens of papers showing that scientists had found the same persistent chemicals everywhere they had bothered to look. The contamination was truly global and well documented.

There was also little doubt that a surprising number of these persistent synthetic chemicals can interfere with hormones and disrupt development. New ones seemed to be cropping up all the time. And if there had been any question about human vulnerability, the work of Howard Bern, John McLachlan, Earl Gray, and others had clearly demonstrated that DES and other estrogen impostors cause the same kinds of damage in most mammals. Given the remarkable similarity in the endocrine system across species, the same was likely to be true with other kinds of hormone disruptors. It would be prudent to assume that whatever happened to animals could also happen to humans.

Taken together, the evidence convinced Colborn that the hormone-disrupting chemicals that now permeated the environment posed a *potential* hazard to humans. But were these ubiquitous chemicals in fact a threat? Were humans exposed to enough to cause harm?

Toxicologists are fond of the axiom that it is the dose that makes the poison. The mere presence of a substance doesn't necessarily produce damage. Even though our body fat and blood testify to our exposure to PCBs, DDT, dioxin, chlordane, and a litany of other persistent chemicals, the amounts we carry are counted in parts per billion or even parts per trillion. Unimaginably small amounts.

Nevertheless, Colborn knew that Fred vom Saal's work showed that even tiny shifts in hormone levels before birth had major consequences for mouse pups. Based on his experiments, ten or twenty parts per *trillion* of natural estrogen are not inconsequential.

But estrogen is a natural hormone and an extremely potent one at that. How much of a synthetic chemical does it take to disrupt hormone levels and do lifelong harm? How much? The question haunted Colborn. She kept searching through the scientific literature, moving from one paper to another, looking for clues. With the passion of a pack rat, she collected every sort of relevant evidence and filed even the smallest tidbit in an ever expanding database on hormone disruption. She had been at it for three years, already, trying to synthesize the far-flung studies from hundreds of researchers in dozens of disciplines into a coherent picture. It was the kind of work that almost never gets done by either the government or the universities because there is no support and no reward for doing it; nobody ever got tenure for analyzing and assessing other people's work. Yet how absurd to spend billions on individual scientific studies but virtually nothing to figure out what they collectively say about the state of the Earth.

Colborn herself had only been able to stay on the trail of the hormone disrupters thanks to a stroke of good luck. As an ecological scientist, John Peterson Myers had been fascinated by Michael Fry's work on DDE and gulls when it came out in the late 1970s, realizing then that its implications were likely to extend far beyond birds. A chance encounter with Colborn in 1988 had renewed this interest and sparked an effort by Myers and Colborn to collaborate. Then in 1990 he moved to become director of the W. Alton Jones Foundation and convinced the board of the private philanthropic trust to establish a senior fellowship for Colborn so that she could focus all her energies on the issue.

As she tried to keep abreast of the latest science on half a dozen fronts, she usually had little time to reflect on the implications of what she was doing. But once in a rare while, alone late at night, she would sit by her apartment window overlooking the illuminated dome of the capitol in Washington and think about what all of the pieces might add up to. The prospects were frightening. What were the long-term effects of these hormone-disrupting chemicals? Were we sabotaging our own fertility as well as that of wildlife? Was it possible that we were unknowingly and invisibly undermining the reproductive future of our children? The thought seemed on the face of it preposterous. How could human fertility be in jeopardy when world population was soaring from five billion toward ten? Maybe she was chasing phantoms.

A few months later, any lingering doubts vanished. At a meeting in Ottawa in the summer of 1990, Colborn happened to hear Richard Peterson of the University of Wisconsin talk about the surprising results his lab had found in new chemical studies. The team at the School of Pharmacy had given dioxin, a compound even more notorious than DDT, to pregnant rats to see how it would affect the development of their male offspring. As they had expected, the dioxin did damage to the male reproductive system if the pups were exposed during a critical period in their prenatal development. What surprised the scientists was how little dioxin it took to do the damage. They hadn't given a large dose or repeated doses, but they saw long-term effects on male pups even when the mother rats had ingested only one dose of an astonishingly tiny amount of dioxin at a critical moment. It had taken just a single hit.

Unlike many laboratory experiments where animals receive doses much higher than those found in the environment, these findings had direct and immediate relevance to the real world. The lowest doses given to the mother rats had been very near to the levels of dioxin and related compounds reported in people in industrialized countries such as the United States, Japan, and those in Europe.

In the world of synthetic chemicals, dioxin has enjoyed the reputation of being the worst of the troublemakers—the most deadly, the most feared, and the most elusive to scientists seeking to unravel the secrets of its toxicity. Lab tests had shown dioxin to be thousands of times more deadly than arsenic to guinea pigs, who died after swallowing only one-millionth of a gram per kilogram of body weight, and the most potent carcinogen ever tested in a number of animal species.

Unlike most other hormone disrupting synthetic chemicals, however, dioxin was not created intentionally. Although some dioxin is released by volcanoes and forest fires, the chemical—known to scientists as 2,3,7,8-TCDD and to the public as "the most toxic chemical on earth"—is for the most part an inadvertent by-product of twentieth-century life, a contaminant created during the manufacture of certain chlorine-containing chemicals such as pesticides and wood preservatives, as well as by bleaching paper with chlorine, incinerating trash containing plastics and paper, and burning fossil fuels. Like DDT and PCBs, dioxin is a fat-loving persistent compound that accumulates in the body. And like other persistent chemicals, it has been detected virtually everywhere—in air, water, soil, sediment, and food.

Although discussion usually focuses on 2,3,7,8-TCDD, it is important to remember this is only the most toxic and notorious member of the dioxin family, which contains 74 other problematic chemicals. Moreover, dioxin is found more often than not in the company of furans—a related family of contaminants containing 135 chemicals with a structure similar to dioxins and with similar toxic and biological effects on animals.

The still ongoing Agent Orange controversy centers on this potent chemical. From 1962 to 1971, the U.S. military dumped more than 19 million gallons of synthetic herbicides over 3.6 million acres in Vietnam in an effort to strip away the rain forest canopy where the U.S. military command believed enemy forces were hiding.

One of the primary weapons in this operation was Agent Orange, the military's name for a mixture that contained the herbicides 2,4-D and 2,4,5-T, the latter a chemical easily contaminated with dioxin during its manufacture. At the height of the campaign against Vietnam's rain forest, enlisted men sprayed Agent Orange not only from airplanes and helicopters but also from boats, jeeps, trucks, and on foot using backpack sprayers.

In the years following their return from Vietnam, veterans reported a variety of personal and family medical problems ranging from cancer to handicaps in their children. As they learned that Agent Orange had been contaminated with dioxin, many became convinced that their own and their children's health problems were linked to this wartime exposure.

After years of debate about whether dioxin was responsible for the reported illnesses, a panel from the National Academy of Sciences at the request of Congress undertook a comprehensive review of the scientific evidence. In their 1993 report, the group found sufficient evidence to link exposure to dioxin-contaminated herbicides to three cancers: soft-tissue sarcoma, non-Hodgkin's lymphoma, and Hodgkin's disease.

In 1979, the U.S. Environmental Protection Agency suspended the use of 2,4,5-T for most purposes, but only after the herbicide had also been used widely on the home front to keep down weeds and brush in suburban lawns, rice fields, pasturelands, and coniferous forests, along highway shoulders and railroad tracks, and under utility power lines. The major nonhousehold uses accounted for almost 7 million pounds of 2,4,5-T in 1974, according to federal figures. Numerous other countries have also banned or withdrawn the legal registration of 2,4,5-T, which prevents legal sale or use, but some nations, such as Australia, have taken no action to restrict its use.

Dioxin gained further notoriety during two dramatic contamination incidents in the United States and Europe.

An explosion at a chemical factory in northern Italy in July 1976 spread a cloud of dioxin over the city of Seveso north of Milan, contaminating almost nine hundred acres of land and thousands of people living nearby. Two weeks after the accident, officials finally decided to evacuate 724 people from the most heavily tainted area.

The dioxin levels measured in some Seveso residents—up to 56,000 parts per trillion—are the highest ever reported in humans.

Though the accident caused no fatalities, scientists are still debating how much it harmed those who were exposed. Shortly after the incident, at least 183 cases of chloracne, a skin disease linked specifically to high exposure to dioxin, were confirmed.

Whether the dioxin exposure caused increased miscarriages and birth defects is unclear. Many pregnant women sought abortions following the explosion, and under any circumstances, it is extremely difficult to detect changes in the rate of miscarriages, since many occur early in pregnancy often without the woman's knowledge that she had been pregnant. Health officials could not confirm the belief that some birth defects had increased after the accident because Seveso had no registry for birth defects before the incident.

While much of the followup research has focused on the possible increased risk of cancer among those exposed in the Seveso accident, insufficient time has passed to fully assess such effects. The preliminary studies examining cancer incidence to date have reported elevated rates for some cancers, but these have been controversial, in part because of the difficulties in accurately determining exposure to those living at varying distances downwind from the plant.

Until recently, no one had thought to survey the children of women exposed to high levels of dioxin in the Seveso accident for any effects save obvious birth defects. Studies are now under way to determine if there have been delayed effects on their sexual development and fertility.

In 1982 and early 1983, Times Beach, Missouri, became a ghost town after the federal government evacuated all 2,240 residents because of dioxin contamination. A company paid to spray the dirt roads to keep the dust down had used a waste oil tainted with dioxin; floodwaters had subsequently spread the contamination into homes and businesses.

Almost a decade earlier, this company, a waste oil hauling operation, experienced a similar incident when it had sprayed the floor of an indoor horse arena with contaminated waste oil. Shortly afterward, according to reports, horses began to sicken and die, and birds that lived in the rafters began dropping to the ground. The owners of

the arena and two young children also became sick with a flulike illness, but while sixty-two horses eventually died, the stricken humans
survived the contamination.

In the Times Beach case, two small studies have looked at the effects on children born to exposed mothers, finding evidence of immune system abnormalities and brain dysfunction, particularly in
the bilateral frontal lobes. The second study focusing on brain effects,
which involved seven boys and girls, found greater dysfunction in the
girls than in the boys, suggesting that the hormonelike activities of
dioxin may have a greater impact on developing females. Researchers
believe that abnormal function in this area of the brain can indirectly
affect thinking processes by altering attention, emotional states, and
motivation.

Few synthetic chemicals have received more scrutiny than
dioxin, in part because of its legendary toxicity. Over the past two
decades, the government and private industry have funded hundreds
of millions of dollars in research to look at everything from how
dioxin acts inside cells to whether workers exposed to high levels on
the job get more cancer. The host of studies have yielded interesting
and sometimes worrisome findings showing that dioxin has a wide
range of effects on the body, such as lowering the sperm count in exposed men and suppressing the immune system. Nevertheless, the
heated public debate in the United States about the dangers of
dioxin focused almost exclusively on whether or not it was, in fact, a
potent carcinogen. Based in large part on a reanalysis of a fourteen-
year-old study of dioxin-induced liver tumors in rats and creative
interpretations of new scientific and epidemiological findings, the
paper industry kept arguing in the late 1980s that dioxin was less
dangerous than previously believed. In 1991, the U.S. Environmental
Protection Agency reevaluated its position on dioxin.

The EPA's reassessment of the risks of dioxin was already under
way when Richard Peterson's Wisconsin study hit with the shock of
an unanticipated asteroid. Here was evidence that dioxin could have
dramatic effects at very low doses—at levels close to those routinely
found in humans. In a matter of months, the tide turned and the
dioxin debate shifted from dioxin's cancer-causing potential to its
developmental and reproductive toxicity. In short order, EPA scien-

tists repeated the studies giving dioxin to pregnant rats and found similar effects in female offspring.

This turnaround in scientific thinking was stunning. The studies suggested that the worst fears about dioxin might, in fact, be justified. Dioxin might after all be more dangerous than anyone had suspected, but contrary to what many had thought, its greatest threat was not cancer. The newly emerging hazard was its power to disrupt natural hormones.

Dioxin's bad reputation helped insure a steady flow of funding to a host of researchers who were probing what this chemical did to the body and how it did it, but the University of Wisconsin lab headed by Peterson was one of the few places exploring its effects on the endocrine system. Robert Moore, one of Peterson's colleagues at the School of Pharmacy and the Environmental Toxicology Center, had set off on this line of research because he believed it held the greatest potential for explaining the toxic effects of the notorious 2,3,7,8 TCDD.

Dioxin posed a fascinating challenge for toxicologists like Moore and Peterson because it is not your ordinary poison. Animals given lethal doses of dioxin don't keel over quickly; they lose their appetite and undergo a mysterious wasting before they actually die weeks later. Dioxin also produces a variety of other nonlethal responses that occasionally seem contradictory. It somehow disrupts estrogen responses, acting sometimes as if it were an estrogen impostor and sometimes as if it were blocking estrogen, yet studies have shown that dioxin is not a simple estrogen mimic like DES. It produces apparently estrogenic or antiestrogenic effects without consorting with the estrogen receptor. For all the years of research, exactly how dioxin does its harm has remained elusive. Peterson and Moore thought the endocrine system might hold the key to this mystery.

As they had suspected, their experiments with adult male rats confirmed that dioxin could interfere with hormone levels. When *adult* rats were given dioxin, it caused their testosterone levels to drop and their testicles and accessory sex organs to lose weight. But it took a lot of dioxin to produce such responses—almost enough to start killing some of the rats used in the experiments.

Although Moore and Peterson felt it was easier to explore the mechanism of toxicity if one used high doses, this approach began to fall into disfavor by the mid-1980s. Critics were attacking high-dose experiments, saying that they did not have direct relevance to the real world where humans and animals are exposed to much smaller amounts of dioxin.

In the end, Moore and Peterson had little choice. The National Institutes of Health, which was funding their work, pushed them toward working with lower doses that the federal agency regarded as more immediately relevant to human health risks. "We got the message that we had to get out of high-dose research if we wanted to stay funded," Moore said.

Well before their high-dose TCDD experiments had ended, Moore read a 1983 paper by Dorothea Sager, a researcher at the University of Wisconsin's Green Bay campus, which found a variety of changes, including reduced fertility, in male rats exposed to PCBs through their mother's milk. Sager's work demonstrated the critical importance of timing, not just in the severity of the impact but also in its very nature. Her findings inspired Moore and Peterson to look for similar patterns as a result of TCDD exposure. They brought in a graduate student, Tom Mably, to conduct the actual experiments.

This team looked beyond the simple question of whether or not rats exposed to dioxin could later produce offspring. This all-or-nothing approach was grossly inadequate. They wanted instead to look at more sensitive aspects of reproductive health, such as sperm counts and mating behaviors that are not often measured in toxicity research. As Moore puts it, "we were looking for answers in different ways. We were turning over different rocks." In fact, Moore reflects, if they had done no more than the usual fertility tests, the work would have sunk into obscurity without a ripple.

Mably's results far exceeded their expectations. While it took an almost lethal dose to impair the reproductive system in adult rats, they found that even small doses did long-term damage to the reproductive system of males exposed in the womb and through their mother's milk. In this study, the mother rats had swallowed only a single dose of dioxin on the fifteenth day of their pregnancy, a criti-

cal period in the process of sexual differentiation that causes males
to become male and not female. As they matured, the male pups
born to mothers given dioxin showed sperm reductions of as much
as fifty-six percent when compared to those whose mothers had ingested none. Moreover, even at the lowest dose the male pups
showed a sperm count drop of as much as forty percent.

"It's a dramatic illustration of how sensitive the male reproductive system can be at a critical stage of development," Moore said. "If
we gave the same dose to a sexually mature male there is nothing we
could detect in terms of reproductive effects." The male reproductive system, they found, is about one hundred times more sensitive
to dioxin during early development than in adulthood.

Dioxin also appeared to affect the sexual behavior of male pups
exposed early in life, suggesting that it had interfered with the sexual
differentiation of the brain. At maturity, these males showed diminished male sexual behavior in mating encounters and increased
propensity to exhibit feminized sexual behavior, such as arching of
the back in a typically female lordosis response, when treated with
hormones and then mounted by another male.

Earl Gray repeated this dioxin experiment at the Environmental Protection Agency's reproductive toxicology lab in Research
Triangle Park, North Carolina, using a different strain of rat and
hamsters, the species considered to be the least sensitive to dioxin.
In toxicity tests, toxicologists have found that the lethal dose for
adult hamsters is one hundred times greater than that which kills
most other animals. Like Peterson, the EPA lab found sharp reductions in male sperm count in rats and hamsters, and in similar studies on female rats, it found malformations of the reproductive tract.
The hamster results were particularly interesting to Gray, since some
have argued that dioxin does not pose a hazard because no humans
have ever died from dioxin exposure. While it might be hard to kill
an adult hamster with dioxin, the species proved as sensitive as the
others to prenatal exposure.

Like Peterson's lab, Gray also found changes in the sexual behavior of male rats, but he is not fully convinced that the diminished
mating prowess is due to altered brain development. There is also
the possibility that dioxin disrupts the development of the male's

genitals so his equipment does not work properly, making him less effective at mating. For the moment, the question of whether dioxin interferes with the development of the brain and thereby disrupts sexual behavior remains unresolved.

Scientists understand less about dioxin than they do about the more straightforward hormone mimics or blockers, such as methoxychlor and vinclozolin, which cause disruption by binding with estrogen or androgen receptors. For this reason, Gray explains, he would be less confident predicting what might happen to humans based on animal experiments. Recent discoveries are, however, giving scientists increasing confidence that the responses in humans and animals are likely to be roughly similar. Researchers have found that dioxin acts almost exclusively through a receptor—one of the "orphan" receptors whose normal chemical messenger remains unknown. Although this receptor was first identified in animals, studies have shown that humans also have a fully functional aryl hydrocarbon, or Ah, receptor that binds to dioxin. Once dioxin occupies the receptor in a human cell, researchers have found it binds to DNA in the cell nucleus, prompting many of the same changes in gene expression seen in animal experiments. Humans seem no less sensitive to this effect. But what happens afterward to produce all of dioxin's disparate biological effects, including developmental disruption, remains a mystery.

However it happens, dioxin acts like a powerful and persistent hormone that is capable of producing lasting effects at very low doses—doses similar to levels found in the human population.

The outstanding irony is that the rats in Moore's experiments passed the standard fertility tests with flying colors—tests typically used by the chemical industry to screen chemicals for safety. Almost all were able to impregnate females and produce the normal number of pups. The reason, Moore explains, is that rats are incredibly robust breeders, producing ten times more sperm than they really need to reproduce. Tests have found that a toxic chemical can knock out ninety-nine percent of a rat's sperm and still have no effect on his ability to reproduce.

Humans, by comparison, are inefficient breeders, who tend to produce barely the number of sperm required for successful fertilization. Moore describes the human male sperm count as "borderline pathological" for many individuals even without the assault of endocrine-disrupting chemicals. If Moore is correct and if long-term declines in human sperm count continue, our species faces a troubling prospect. Such a drop could have a devastating impact on human fertility.

HERE, THERE, AND EVERYWHERE

THE THREAT OF HORMONE-DISRUPTING CHEMICALS HAS COME TO LIGHT largely through a series of accidental discoveries and surprises, but none more bizarre than the incident that began just after Christmas in 1987 at Tufts Medical School in Boston.

On the sixth floor of an old brick building on the edge of Chinatown, Dr. Ana Soto donned her white coat and headed for her lab to check out the cell cultures in the latest round of experiments, expecting that the day would proceed according to the usual steady, careful routine. For more than two decades, she and her fellow physician Carlos Sonnenschein have been exploring why cells multiply—a fundamental question in basic biology as well as one central to the mystery of cancer, where cell multiplication runs amok.

The pair, who first teamed up in 1973 when Soto joined Sonnenschein's laboratory at Tufts, had been working on a theory that challenged the conventional wisdom on cell proliferation. The prevailing theory holds that something in the body, which scientists have dubbed "growth factors," prompts cells to multiply, a view that assumes nonproliferation is the normal state.

In the beginning, Sonnenschein and Soto had been looking for growth factors, too, but puzzling results in their experiments had led them to reexamine their assumptions. In tackling the question from an evolutionary perspective, Sonnenschein and Soto had come to think the opposite was likely to be true. After all, they reasoned, the single-cell organisms that first evolved, such as bacteria, do not need anything to tell them to grow; they reproduce endlessly if the food and conditions are right. Well-nourished cells only stop continuous proliferation when they become a part of a multicellular organism, suggesting that some inhibiting substance keeps them in check.

The question probably wasn't what makes cells multiply; it's what makes them stop. The complex organisms that appeared later in the history of life cannot survive unless their cells maintain a certain discipline. If these cells start multiplying continuously in the way bacteria do, the organism will rapidly turn into little more than one big disorganized tumor. For Soto and Sonnenschein, the goal was to find such an inhibitor.

They were in hot pursuit of this inhibitor through experiments with human breast cancer cells—a strain that multiplies in the presence of estrogen. Under normal circumstances, estrogen prompts tissue growth in the breast and uterus—something Soto and Sonnenschein think the hormone does by overriding the inhibitor. The hormone has a similar effect on the estrogen-responsive line of cells that the pair use in their research. When estrogen is added to these living cells in a lab dish, the cells will multiply. The pair was confident this cell culture research could help them track down the inhibitor.

By 1985, Sonnenschein and Soto had found evidence that their proposed inhibitor really did exist. If they removed the estrogen from blood serum through a special charcoal filtering process and then added the serum to estrogen-sensitive breast cancer cells, the cells would stop multiplying. Two years later, they were struggling to isolate and purify the specific substance in the serum that had given the stop signal.

Working with cells in tissue culture can be a tricky business. There is only one way of doing things—impeccably. Any lapse in discipline, the least hint of sloppiness, can ruin weeks, months, even years of work. To eliminate potential problems, Sonnenschein and

Soto ran the lab with very few people and followed carefully defined procedures to maintain maximum control. They kept the hormones used in the experiments locked in a tackle box in a different lab. Any work with the cells, they did themselves. Their system of elaborate precautions bordering on paranoia had paid off. They had never had a problem—until that final week in 1987.

Four days earlier, Sonnenschein had prepared a series of multi-welled plastic plates, placing breast cancer cells in the twelve small cups and then adding varying levels of estrogen and of the estrogen-free serum to each of the tiny cell colonies. Now Sonnenschein and Soto were returning to see how the cells had fared. Over the years, they had done variations of this experiment hundreds of times. According to the routine, they would examine the cells under the microscope before transferring the cells from the plates to special counting vials, so the cells could be tallied by an electronic particle counter.

Somehow the plate didn't look right, so Sonnenschein adjusted the microscope and looked again. His eyes were not playing tricks. The whole plate—every single colony growing in a specially modified blood serum—was as crowded as a subway train at rush hour. Regardless of whether they had added estrogen or not, the breast cancer cells had been multiplying like crazy.

In all their years of cell work, they had never seen anything like it. At first, they felt stunned. They didn't know what to think except that something had gone seriously wrong.

It had to be some sort of estrogen contamination, they presumed. They could see that immediately because Carlos was working with other cells as well and these were behaving as expected. The only cells multiplying wildly were the estrogen-sensitive breast cancer cells.

They carefully prepared another batch of plates with breast cancer cells, and once again, they saw the same galloping proliferation. It wasn't a fleeting event. The mysterious contamination was still somewhere in the lab.

Soto and Sonnenschein spent the New Year's holiday fighting depression and going over their lab procedures again and again, searching for changes or possible slipups that might account for the runaway proliferation. Since they did all the cell work themselves, there was no one else to blame. But what could they possibly have done wrong? And if they hadn't slipped up, what else could have happened?

Maybe the breast cancer cells had somehow changed or been contaminated by foreign cells.

No, they quickly ruled that out by comparing the cells with frozen samples of the same cell line and with other estrogen-responsive cells. When tested, all showed the same mysterious proliferation without exposure to estrogen.

They considered every possible explanation from carelessness to sabotage. Had somebody unthinkingly entered the lab with an open bottle of estradiol, the form of estrogen used in their experiments? The hormone is so powerful they keep only one gram on hand for their research—less than a teaspoon. A stray speck could contaminate a lab. That's why it was locked away in another lab. Given the tight controls, such an accident seemed highly unlikely.

Could someone have derailed their experiments intentionally? That thought crossed their minds when they failed to come up with more mundane explanations. Sabotage of experiments by jealous colleagues was not unheard of in science. Their work was breaking new ground and upsetting long-held notions. But there had to be a more reasonable explanation.

In the end, the cause proved beyond their wildest imaginings, something even stranger and more unsettling than human sabotage. It would be four long, frustrating months before they finally tracked down the "phantom estrogen" and two whole years before they were able to put a name to the chemical that was mimicking the hormone.

Their discovery shook even veteran investigators of hormone-disrupting chemicals. For years, the ongoing discussion about possible human health risks from synthetic chemicals had been based on the assumption that most human exposure comes from chemical residues, primarily pesticides, in food and water. Now Soto and Sonnenschein had discovered hormone-disrupting chemicals where you would least expect them—in ubiquitous products considered benign and inert. Here was glaring evidence of our vast ignorance about hormone-disrupting chemicals in the environment and how we might be exposed to them.

* * *

With the new year, Sonnenschein and Soto began their detective work in earnest. Until they tracked down the contamination, their research was stopped cold.

Having no clues whatsoever, they finally decided to attack the mystery through a tedious process of elimination. They started by making an exhaustive list detailing each step in their experiments and every piece of equipment used. Then they set about repeating the experiment, replacing a single item at each run-through. Maybe some technician had failed to give the glass pipettes a proper acid washing to remove all traces of estrogen. They used a brand-new pipette to see if that made a difference. What about the charcoal? Would using charcoal from a new bottle to strip the estrogen from the serum alter the outcome? How about the plastic tubes where the serum is stored? They tried a new batch.

Despite their precautions, nothing they tried made any difference. Time after time, their hearts sank as they saw dense masses of cells through the microscope. With every day and every failure, they fell further and further behind in their research schedule. The contamination haunted them. It got to the point where they did not do anything but watch the cells to see if they had finally found the source of the problem.

Sonnenschein began to suspect that the problem must be in the lab itself. Somehow the very atmosphere must be contaminated. To test this latest suspicion, he arranged to use a colleague's lab and all its equipment. The only thing he carried into this lab were the breast cancer cells and the tubes of charcoal-treated serum. Foiled again. The slides still showed signs of rampant cell growth.

That meant the contamination had come with them—in the charcoal or the tubes. But how could that be, since they had checked out the charcoal and they had been using the same lab tubes for years?

Wearily, they prepared to conduct the experiment yet another time, but in this run they would try a different brand of lab tube— one made by Falcon instead of Corning. By now, it was the end of April. They had been on the hunt for the contaminant for four long, maddening, depressing months. The tension had become almost unbearable.

As Sonnenschein prepared to examine the plates several days

later, Soto and others gathered around, watching silently as he slipped one under the microscope.

"They're inhibited," he announced triumphantly.

At last. Soto felt the tightness that had gripped her ebb away.

This had to be it. The orange-capped Corning tubes. Something must be leaching from the tubes into the blood serum they used in their experiments, something that acted like estrogen. The plastic, which they had always regarded as a benign, inert substance must contain chemicals that can cause significant, worrisome changes in human cells. Far from inert, the plastic appeared biologically active.

Their elation at having solved the puzzle proved short-lived, however, for they were quickly overtaken by a growing sense of foreboding. If this was happening with their lab tubes, it must be happening with other plastic products as well. This was almost certainly a problem that extended far beyond their lab.

Within days, Sonnenschein contacted Jean Mayer, then president of Tufts, who was a nutritionist by training. He immediately grasped the implications as well. University officials then alerted Corning's scientific products division about the problem. The company responded by providing a set of coded test tubes for another round of experiments. The results were curious indeed. When they stored the hormone-free blood serum in some of the tubes, their breast cancer cells showed an estrogenlike reponse. But the cells showed no response to serum stored in other identical-looking tubes. The findings prompted Corning officials to meet with Sonnenschein and Soto and other Tufts representatives on July 12, 1988.

At that meeting in the Hilton at Boston's Logan Airport, Soto and Sonnenschein learned that the company had recently changed the plastic resin to make the tubes less brittle but had not bothered to change the catalog number. Although the medical school lab kept ordering the tube number they had used for years, Corning was now supplying a lab tube that had a different chemical composition. When Soto asked about the chemical content of the new resin, Corning declined to disclose the information on the grounds that it was a "trade secret."

Soto and Sonnenschein were outraged. What were they to do? Simply switch to another brand of tube and go on with their own research? Could they in good conscience ignore the implications of their accidental discovery?

Giving up would have been not only irresponsible but also out of character. As a woman from a Latin country in a profession dominated by men, Soto had not gotten as far as she had by taking no for an answer. Such a refusal simply brought out her terrierlike tenacity. Although Carlos might appear more easygoing, he shared a similar stubborn streak. It was evident in their research. Taking on the establishment in one's field requires a certain independence and determination.

Though they were cell biologists by training, not organic chemists, and had no money for this kind of investigation, one way or another, they agreed, they would find out what the chemical was that was leaching out of the plastic and causing rampant proliferation in their breast cancer cells.

Soto wondered what effect tubes of this same composition would have if they were used in diagnostic tests in laboratories. Sonnenschein thought about all the plastics used to package food, even in baby bottles. Sonnenschein worried that kids might be taking in estrogenic substances with their milk. As physicians and researchers who had spent decades studying hormone effects, they held the strong conviction that increasing estrogen exposure is risky and unwise.

It took months to purify the compound in the plastic that caused the estrogenlike effect in their experiments and do a preliminary identification using mass spectrometry analysis. Finally, they were ready to send a sample of the substance across the river to chemists at the Massachusetts Institute of Technology for final identification.

At the end of 1989—two years after their detective work had started—they had a definitive answer: p-nonylphenol.

Through further investigation, Soto and Sonnenschein learned that p-nonylphenol is one of a family of synthetic chemicals known as alkylphenols. Manufacturers add nonylphenols to polystyrene and polyvinyl chloride, known commonly as PVC, as an antioxidant to make these plastics more stable and less breakable. The plastic centrifuge tubes in which they stored blood serum had been of polystyrene—a plastic that, depending on the manufacturer, may or may not include nonylphenols.

In a search of the scientific literature, they found bits and pieces of information that only heightened their concern. One study had found that the food processing and packaging industry used PVCs that contained alkylphenols. Another reported finding nonylphenol contamination in water that had passed through PVC tubing. Soto and Sonnenschein even discovered that nonylphenol is used to synthesize a compound found in contraceptive creams—nonoxynol-9. In studies with rats, researchers had found that the nonoxynol-9 breaks down once inside the animal's body, creating nonylphenol.

They also learned that the breakdown of chemicals found in industrial detergents, pesticides, and personal care products can likewise give rise to nonylphenol. The United States and other countries use vast quantities of these chemicals called alkylphenol polyethoxylates—450 million pounds in 1990 in the United States alone and more than 600 million pounds globally. Although the products purchased by the consumer, such as detergents, are not themselves estrogenic, studies have found that bacteria in animals' bodies, in the environment, or in sewage treatment plants degrade these alkylphenol polyethoxylates, creating nonylphenol and other chemicals that do mimic estrogens.

In follow-up work, Soto and Sonnenschein took the chemical they had found in their lab tubes and injected it into rats to verify that it acted like estrogen in living animals as well as in cells in a lab dish. In tests with female rats without ovaries, they found that p-nonylphenol would cause the lining of the uterus to proliferate as if the rats had been given estrogen. Because the synthetic chemical is less potent than natural estrogen, it took higher doses to produce an effect.

Alkylphenol polyethoxylates have been widely used since the 1940s, but in the past decade they have come under increasing scrutiny because of their toxicity to aquatic life, particularly as they break down. By the late 1980s, several European countries had already banned the use in household cleaners of nonylphenol ethoxylates, the compound in this group most commonly used in cleaning products, and similar restrictions are under consideration in other countries as well. While many still allow their use, however, in cleaners prepared for industrial purposes, fourteen European and Scandinavian countries agreed in 1992 to phase out this use by 2000.

When Soto and Sonnenschein published their findings in 1991, they added a new concern to the growing list. This was the first report that these widely used and reasonably well studied chemicals might also act to disrupt hormones.

By strange coincidence, while Soto and Sonnenschein were chasing contamination in their lab, a similar drama was unfolding at the opposite end of the country at Stanford University School of Medicine in Palo Alto, California. In this case, too, the mystery estrogen was traced to plastic lab equipment but not to polystyrene products or to nonylphenol. The Stanford team found another estrogen mimic, bisphenol-A, which was leaching from an entirely different kind of plastic, polycarbonate. This plastic is used for lab flasks and for many consumer products such as the giant jugs used to bottle drinking water.

Here again, the discovery was accidental and one that occurred only because the scientists were conducting research with estrogen-sensitive cells. David Feldman, a professor of medicine, and his colleagues in the endocrinology division had initially discovered a protein in yeast that binds with estrogen, which they thought might be a primitive estrogen receptor, and if yeast had such an estrogen receptor, then there must be a yeast hormone. The team was hunting for such a hormone when they saw that some substance was indeed binding to the yeast receptor. But the researchers soon realized the estrogenic effect was due to a contaminant rather than a hormone. They determined that the contaminant was bisphenol-A and that the source of the contamination was the polycarbonate lab flasks used to sterilize the water used in the experiments.

In a 1993 paper, the Stanford team reported their discovery and their discussions with the manufacturer of polycarbonate, GE Plastics Company. Apparently aware that polycarbonate will leach, particularly if exposed to high temperatures and caustic cleaners, the company had developed a special washing regimen that they thought had eliminated the problem. In working with the company, however, the researchers discovered that GE could not detect bisphenol-A in samples sent by the Stanford lab—samples that were causing proliferation in estrogen-responsive breast cancer cells. The problem proved to be the detection limit in GE's chemical assay—a limit of ten parts per billion. The Stanford team found that two to five parts per billion of bisphenol-A was enough to prompt an estrogenic response in cells in the lab.

Though bisphenol-A is two thousand times less potent than estrogen, notes Feldman, "it still has activity in the parts per billion range." Feldman is cautious, however, about making people alarmed about plastics. "We don't know enough yet to make this into a public health crisis." He adds, however, that the accidental discovery about polycarbonate raises a host of questions that need to be answered. The Stanford paper shows that bisphenol-A prompts an estrogen response in cells in a lab. The next logical question, he says, is whether it prompts the same response when given in water to an animal.

Unfortunately, such questions remain unanswered at this time because researchers like Sonnenschein and Soto have been unable to secure the funding to further investigate biologically active plastics and other hormone-disrupting synthetic chemicals. The problem seems to stem largely from the inertia of institutions and ideas. Those seeking to do this research complain that grant reviewers tend to be locked into older ideas, which focus on the damage done by toxic chemicals to DNA. As a consequence, they often don't fully grasp or appreciate the importance of this new line of research and tend to have a narrow view of the types of studies that should receive federal funds. To make matters worse, few if any grant-making institutions are receptive to or prepared to review proposals for the kind of interdisciplinary research that is needed to investigate these questions.

In this same period, John Sumpter, a scientist in a very different field, had been enlisted to help solve another mystery on the other side of the Atlantic—the case of the sexually confused fish. Sumpter is a biologist from Brunel University in Uxbridge, who has studied the role of hormones in fish reproduction.

Anglers fishing in English rivers had been reporting that something strange was happening to the fish, particularly in the lagoons just below the discharges from sewage treatment plants. The problem was not the usual fish kills that can occur because of pesticides or low oxygen levels. Nor did the fish appear to have any obvious disease. But many looked quite bizarre. Even experienced fishermen could often not tell if a fish was male or female, for they showed male and female sexual characteristics at the same time. They seemed perfect examples of the condition scientists refer to as "intersex," where an individual is stranded between sexes.

The government fisheries staff approached Sumpter because they suspected that something in the water—either hormones or a substance that mimics hormones—was causing the sexual confusion. So they put this question to him: is there anything to measure in a fish that will indicate whether it is being exposed to hormones?

It depends on the hormone, Sumpter replied. If the water contained anything that acted like estrogen, he was sure there would be a telltale sign in males. They would respond to the estrogen by making a special egg-yolk protein normally produced only by females. In the females, the liver produces this protein, vitellogenin, in response to an estrogen signal from the ovaries. Once the liver synthesizes vitellogenin, the blood carries it back to the ovaries, where it is taken up and incorporated into the eggs as the female prepares for reproduction. Although males do not produce eggs, their livers will nevertheless produce vitellogenin if they are exposed to elevated levels of estrogen. Since this response is extremely dependent on estrogen, vitellogenin levels found in male fish provide a good indication of estrogen exposure.

The initial question—whether the sewage treatment plants were releasing something that acted like estrogen—proved the easiest to answer. The fish bore unequivocal testimony that this was the case.

Within a week after the research team placed cages of rainbow trout bred in captivity in the water flowing from a sewage treatment plant on the River Lea fifty miles north of London, the vitellogenin levels measured in the blood of the fish soared. The caged fish were producing five hundred times more vitellogenin than trout maintained in clean water elsewhere. In three weeks, levels of Sumpter's telltale estrogen marker climbed even higher, to more than one thousandfold.

A nationwide survey followed in the summer of 1988 involving twenty-eight sites in England and Wales. Although low oxygen levels and plant malfunctions killed off some of the caged fish, the researchers still obtained results from fifteen of these sites, finding dramatic increases in vitellogenin levels in each and every case. Some of the increases were truly staggering: in one fish vitellogenin reached one hundred thousandfold above normal. Whatever this estrogenic substance was, the survey showed it was a national, not a local problem.

The findings raised troubling questions about possible human exposure through drinking water, which is taken for the most part from these rivers. In summer, up to fifty percent of the water in rivers can be effluent from sewage treatment plants, and in a dry summer, the proportion of effluent can rise to as much as ninety percent. Tests with caged fish at eight water reservoirs on rivers, however, found no vitellogenin effects in fish. While this is somewhat reassuring, it shows only that there is not enough of the estrogenic substance present to produce a response in adult fish. They did not test fish during sensitive developmental stages, nor did they look for other, non-estrogen-based hormone disruption. It does not rule out the presence of estrogenic chemicals in drinking water. British water companies are not required by law to test routinely for these chemicals. Even if tests are conducted there is no requirement that they disclose results to the public.

But where the estrogenic substance was coming from proved harder to answer. The first thought was birth control pills. Sumpter and his colleagues theorized that women taking oral contraceptives, which contained a form of estrogen called ethynylestradiol, would excrete it in their urine, so the hormone would eventually end up at the treatment plant and ultimately in the rivers. In laboratory experiments, they established that this form of estrogen does produce effects in fish at concentrations of one-billionth of a gram per liter of water. As hard as they looked, however, the British scientists could not detect this chemical in the water released from sewage treatment plants.

Sumpter and his colleagues then considered other estrogenic substances, such as plant estrogens and pesticides. While these certainly might contribute to some overall estrogenic effect, they thought it unlikely that there was enough coming out of treatment plants to account for any major part of the estrogenic phenomenon they were seeing.

Then, in a search of the scientific literature on estrogenic chemicals, Sumpter happened on Soto and Sonnenschein's paper describing their discovery of nonylphenol in polystyrene plastic and other products. He learned from it that detergents can contain ingredients that degrade into estrogenic chemicals. Here was a new suspect.

The researchers developed a new round of tests to see whether this theory was indeed plausible. First, were the alkylphenols estrogenic to fish as well as to the human breast cancer cells used by the Tufts researchers? And second, is there enough of the substance in the environment to produce an effect in fish? The answer to both questions proved to be yes. The fish did respond, and the levels found in river water were high enough to cause male fish to produce significant amounts of vitellogenin. Studies exposing maturing male fish to concentrations of nonylphenol showed that modest concentrations that can be found in the environment are enough to inhibit the growth of their testicles.

Sumpter and his colleagues now had a strong suspect—the alkylphenols that result from the breakdown of detergents containing alkylphenol polyethoxylates, but they are not yet certain whether this family of chemicals is the culprit. They are now trying to pin this down through chemical analysis. Preliminary data are already showing that the estrogen problem won't be the result of a single chemical but rather of a mixture. "We don't know if the mixture is related to alkylphenol polyethoxylates," Sumpter says. "It could be a single family of chemicals or a mixture of detergents, pesticides, and plasticizers." Sumpter is himself betting it will be "a range of chemicals that all contribute to the effect."

The initial discoveries of hormone-disrupting chemicals in some unexpected places has inevitably led to others.

Spurred by the Tufts researchers' report of biologically active plastics, Spanish scientists at the University of Granada decided to investigate the plastic coatings that manufacturers use to line metal cans. These often inconspicuous coatings had been added because of concerns that metals might contaminate the food or impart a metallic taste. Such plastic linings are reportedly found in eighty-five percent of the food cans in the United States and about forty percent of those sold in Spain.

The brother and sister team of Fátima Olea, a food toxicologist, and Nicolás Olea, a physician specializing in endocrine cancers, had visited the United States and worked at the Tufts Medical School lab with Soto and Sonnenschein. Their time in Boston had alerted them to the potential hazards from plastics.

Their suspicions proved well-founded. In a study analyzing twenty brands of canned foods purchased in the United States and in Spain, they not only discovered bisphenol-A, the same chemical that Stanford researchers had found leaching from polycarbonate lab flasks, they also found stunningly high concentrations in such products as corn, artichokes, and peas. Bisphenol-A contamination was detected in about half the canned foods they analyzed. In some instances, the cans contained as much as eighty parts per billion—twenty-seven times *more* than the amount that the Stanford team reported was enough to make breast cancer cells proliferate. At such levels, a synthetic estrogen mimic might contribute significantly to a person's exposure regardless of whether it is a "weak" estrogen or not.

Biologically active plastics were leaching from cans, containers where one would not expect to find plastic at all.

All the incidents discussed in this chapter involve synthetic chemicals that mimic estrogen, but the synthetic chemicals that disrupt hormones do not all act like estrogens. Other hormones in the body are vulnerable as well. Recall, for example, that some fungicides interfere with the action of male hormones. Moreover, in ongoing studies at the U.S. Environmental Protection Agency lab in Research Triangle Park, North Carolina, Earl Gray has discovered that even the classic estrogen mimics have far broader effects than scientists in this field had hitherto recognized. Some of these "estrogenic" synthetic chemicals, it turns out, also take a direct toll on males by blocking the androgen receptors that respond to male hormones.

The question of exposure lies at the heart of the debate about whether hormone-disrupting chemicals pose a hazard or not.

Some skeptics dismiss such concerns, arguing that the hormone effects of synthetic chemicals are far weaker than those of natural hormones and that humans are not being exposed to enough to pose a hazard.

Such assertions are not supported by the evidence. When one surveys the available information and scientific literature, one quickly discovers that there are far too many blank spaces and missing pieces to provide even a rough picture of how much humans might be taking in or to allow for definitive conclusions. Often the needed information simply does not exist or it is unavailable. Manufacturers frequently withhold information about the ingredients in their products using the claim of proprietary information or trade secrets—a principle that is far more rigorously protected by legal precedent and the courts than is the public's right to know. Even the federal Freedom of Information Act, which is supposed to give U.S. citizens access to information held by the government, contains an exemption for trade secrets or confidential business information. It is anybody's guess how many of the plastic consumer goods on the market contain hormone-disrupting chemicals. Even in the case of pesticides, where governments maintain closer oversight, it is impossible to obtain coherent data on the production of specific pesticides. The government figures available in the United States and elsewhere are limited and disjointed at best.

The information in the scientific literature about the biological activity of and human exposure to chemicals of concern is equally fragmentary and unsatisfactory. In some instances, there are isolated studies reporting on concentrations of one chemical or another in human blood or body fat, describing various sources of exposure, or detailing how a hormone-disrupting chemical affects the liver, the cells, the nervous system, the brain, or some other part of the body.

Oftentimes, the studies show trends that differ from place to place or among chemicals. This is not surprising. The industrialized countries that controlled persistent chemicals such as DDT witnessed a dramatic drop in DDT concentrations in the late 1970s. Although the rates of DDT reduction in human tissue have slowed markedly since then, this is nevertheless an encouraging sign that government action can lower the level of exposure. In developing countries in Latin America, Africa, and tropical Asia, where heavy use of DDT and lindane continues, human tissue still shows significant concentrations of these persistent chemicals.

PCBs are another story altogether. The concentrations in human tissue have remained steady in recent years even though most industrial countries stopped PCB production more than a decade ago. This no doubt reflects the fact that exposure continues despite the ban, because two-thirds of the PCBs ever produced are still in use in transformers or other electrical equipment and therefore subject to accidental release.

One is left with isolated snapshots of one aspect of the problem or another. But in the end, we do not know how to tally these disconnected facts. We do not know what they collectively mean. If anything, it is becoming even harder to assess exposure as persistent chemicals have been phased out in industrial countries and replaced by less persistent compounds such as methoxychlor. Like DDT, methoxychlor disrupts hormones, but unlike its predecessors, it does not leave telltale signs of exposure in body tissue. And the troubling question looms of how many more hormone-disrupting chemicals remain to be discovered.

Although synthetic chemicals now seem an inextricable part of the fabric of modern life, they have come into common use relatively recently. The synthetic chemical industry first developed in the second half of the nineteenth century, after chemists learned to synthesize textile dyes in the laboratory and manufacturing of these man-made dyes began on a large scale. But the "chemical age" that has transformed daily life did not dawn until around World War II, when new discoveries and new techniques revolutionized the industry and led to an era of explosive expansion in the production of synthetic chemicals. Between 1940 and 1982, production of synthetic materials increased roughly 350 times, and billions of pounds of man-made chemicals poured into the environment, exposing humans, wildlife, and the planetary system to countless compounds never before encountered.

Consider a few figures that sketch the magnitude of this global experiment that has been ongoing for half a century now.

U.S. production of carbon-based synthetic chemicals, which represent the lion's share of synthetic chemicals, topped 435 billion pounds in 1992, or 1,600 pounds per capita. Global production is estimated to be roughly four times greater, but actual figures are impossible to come by.

Around the world, one hundred thousand synthetic chemicals are now on the market. Each year one thousand new substances are introduced, most of them without adequate testing and review. At best, existing testing facilities worldwide can test only five hundred substances a year. In reality, only a fraction of this number actually do get tested.

The world market in pesticides amounted to 5 billion pounds in 1989 and included sixteen hundred chemicals. Worldwide use is still increasing. Pesticides are a special class of chemicals in that they are biologically active by design and intentionally dispersed into the environment.

Today the United States uses thirty times more synthetic pesticides than in 1945. In this same period, the killing power per pound of the chemicals used by 900,000 farms and 69 million households has increased tenfold. Pesticide use in the United States alone amounts to 2.2 billion pounds a year, roughly 8.8 pounds per capita.

Thirty-five percent of the food consumed in the United States has detectable pesticide residues. U.S. analytical methods, however, detect only one-third of the more than six hundred pesticides in use. Pesticide contamination of food is often much higher in developing nations. In Egypt, most milk samples surveyed in one study contained high residue levels of fifteen pesticides.

The world trade in chemicals includes fifteen thousand synthetic chlorinated compounds—a category of chemicals that has come under attack because of their persistence and a record of causing health and environmental problems. Although most industrial countries imposed restrictions in the 1970s on the most notorious chemicals in this class, in developing countries, where they are used to control pests that threaten public health and crops, their use is increasing.

In 1991, the United States exported at least 4.1 million pounds of pesticides that had been banned, canceled, or voluntarily suspended for use in the United States, including 96 tons of DDT. These exports included 40 million pounds of compounds known to be endocrine disruptors.

The amounts of potentially harmful chemical compounds produced each year are truly staggering—thousands upon thousands of
chemicals and billions of pounds. Five billion pounds of pesticides
alone are spread far and wide not only on agricultural fields but in
parks, schools, restaurants, supermarkets, homes, and gardens.

At best, a few hundred of these chemicals have been studied in
any depth or detail—among them the persistent compounds such as
DDT, PCBs, dioxin, and lindane. But even here, it appears that our
ignorance far exceeds our knowledge. Despite the billions of dollars
that have been invested in dioxin research, the recent findings at the
University of Wisconsin—showing that very low doses can have profound long-term effects on the reproductive system of those exposed
in the womb—took most by surprise, including the scientists conducting the experiment.

Few if any safety data exist for many of these chemicals. The
safety data that do exist are typically limited to whether the chemical
may cause cancer or gross birth defects. Possible effects on the endocrine system or transgenerational effects are rarely, if ever, examined.

The existing information may not allow any reliable estimates
regarding human exposure to hormone-disrupting chemicals and the
magnitude of the hazard, but there is enough evidence to raise profound and troubling questions. As scientists have begun to explore
the possible threat, new discoveries have only heightened concern.
Just as leading researchers had predicted, more endocrine-disrupting
chemicals have been discovered as the first systematic investigations
have gotten under way. And more are expected.

Contrary to assertions by critics, the hormonal activity of synthetic chemicals is not always "weak." In recent studies, Earl Gray
has found that p'p-DDE, a degraded form of DDT that is ubiquitous
in human body fat, is a potent androgen blocker and equal in power
to flutamide, a drug designed to block the receptors for male hormones that is used in treating prostate cancer.

The discovery that hormone-disrupting chemicals may lurk in
unexpected places, including products considered biologically inert
such as plastics, has challenged traditional notions about exposure
and suggests that humans may be exposed to far more than previously believed.

Even more worrisome, scientists are now finding evidence that hormone-disrupting chemicals can act *together* and that small, seemingly insignificant quantities of individual chemicals can have a major cumulative effect. Ana Soto and Carlos Sonnenschein have now demonstrated this with breast cancer cells in culture. When they exposed the estrogen-sensitive breast cancer cells individually to small quantities of ten chemicals known to be estrogen mimics, they found no significant growth in the cells. But the cells showed pronounced proliferation when these same small quantities of the same ten chemicals were given together. Sumpter is finding evidence of additive effects as well.

Scientists have also been exploring the critical question of whether special blood proteins bind to synthetic estrogen mimics in the way they do to estrogen. In pregnant women, this binding action ties up most of the estrogen circulating in the blood and acts to protect fetuses from excessive hormone exposure in the womb. As studies proceed, scientists are finding increasing evidence that the blood proteins do not bind to synthetic chemicals just as they do not bind to DES. If all of a synthetic mimic is free and unbound, this would greatly increase the potential disruption and increase concern about relatively small doses of so-called weak estrogens or other hormone disruptors.

Each and every discovery discussed in this chapter has added to scientific knowledge about endocrine-disrupting chemicals, but ironically, these discoveries have also underscored our astonishing ignorance about the man-made chemicals that we have spread liberally across the face of the Earth and incorporated into every part of our daily lives.

In truth, no one yet knows how much it takes of these synthetic hormone-disrupting chemicals to pose a hazard to humans. All evidence suggests that it may take very little if the exposure occurs before birth. In the case of dioxin at least, the recent studies have shown that human exposure is sufficient to be of concern.

Despite our vast ignorance, we should not lose sight of some important things we do know.

Most of us carry several hundred persistent chemicals in our body, including many that have been identified as hormone disruptors.

Moreover, we carry them at concentrations several thousand times higher than the natural levels of free estrogen—the estrogen that is not bound up by blood proteins and is therefore biologically active.

As Fred vom Saal has discovered, vanishingly small amounts of free estrogen are capable of altering the course of development in the womb—as little as one-tenth of a part per trillion. Given this exquisite sensitivity, even small amounts of a weak estrogen mimic—a chemical that is one thousand times less potent than the estradiol made by the body itself—may nevertheless spell big trouble.

CHRONICLE OF LOSS

THE HYDRAULIC CRANE GROANED AS IT HOISTED ITS PALE BURDEN OUT of the water. For a brief moment, the one-ton beluga whale seemed to swim in the air as the crane swung toward the long, thin trailer waiting on the dock at Mont-Joli, on the south bank of the St. Lawrence River in Quebec. It deposited the whale with a dull thud.

Earlier that day, May 31, 1989, a fisherman had found the animal floating belly up offshore and had towed it to the jetty at Pointe-aux-Cenelles. Out on the river, belugas are dazzling creatures—all magic and grace when one catches sight of them slipping through the waves like watery angels. Yanked from their element, they look different indeed. If it resembled anything, the thirteen-foot-long whale carcass parked on the trailer looked like a huge bloated sausage, except that it was strikingly white. Like porcelain.

"Another entry for the Book of the Dead," Pierre Béland thought as he turned and climbed into the cab of the high-powered diesel Ford pickup to start the long drive back toward Montreal. Over the past seven years, Béland, a scientist and founder of the St. Lawrence National Institute of Ecotoxicology, had logged thousands of kilometers driving back and forth along both banks of the St. Lawrence to fetch dead belugas and bring them back to the veterinary school at the University of Montreal. In six hours or so, he would be back in the autopsy room with his colleagues, veterinarian Sylvain De Guise and technician Richard Plante, digging into the bowels of another whale in the middle of the night. The findings would add to their growing record on St. Lawrence belugas— Béland's Book of the Dead, a chronicle of loss as well as a valuable body of scientific data.

By now, Béland and his colleagues figured they had seen just about every kind of abnormality, but the whale hurtling down the highway behind him would prove no routine case. This beluga harbored a freakish secret that would earn it a special place in the annals of science.

For years, scientists had offered theories about why the St Lawrence beluga population had continued to decline even after large-scale commercial whaling had ceased in the 1950s, blaming the dwindling numbers on overexploitation and a subsequent loss of habitat through dredging and hydroelectric projects. If the pollution in the St. Lawrence was mentioned at all, it was treated as a minor factor. Whaling, without question, had taken a devastating toll. The population had plummeted from an estimated five thousand at the turn of the century to twelve hundred in the early 1960s. But in the three decades since then, beluga numbers had continued downward to a current population of about five hundred.

Béland entered the debate and the field of whale research quite by accident. Although trained as a biologist, he had spent more time with computers than with animals in the field, earning his Ph.D. in mathematical ecology, which seeks to probe the dynamics of an ecosystem through mathematical models and equations. In September 1982, he was working in the marine ecosystems branch of a new federal Fisheries and Oceans research center in Rimouski on the south shore of the St. Lawrence when a beluga washed up nearby. Out of curiosity, he had gone out with a young veterinarian, Daniel Martineau, to take a look at the beluga, an animal he had seen only at a distance as a dancing white flash in the broad expanse of blue gray water. As the two men were standing there on the beach, Martineau, who was far more interested in the whales he saw on the river than in the cows he routinely treated, suddenly suggested they cut it open to try to find out why it had died.

Even that first encounter suggested that synthetic chemicals might be a greater factor in the beluga decline than anyone had recognized. The Fisheries and Oceans lab in Montreal, where Béland had sent a sample of the whale's tissue, reported that the animal was highly contaminated with toxic chemicals, including DDT, PCBs, and mercury. When two more belugas washed up that fall, Béland and Martineau examined them as well and found a variety of abnormalities and lesions never before reported in whales.

In the intervening years, Béland, working with Martineau and De Guise, had taken part in dozens of whale autopsies and compiled an astonishing list of afflictions, most of them never seen in beluga populations inhabiting less polluted waters. The St. Lawrence whales had malignant tumors, benign tumors, breast tumors, and abdominal masses. One had bladder cancer like many of the workers at the aluminum plant on the Saguenay River, a tributary where some of the whales spend a good deal of time. They suffered from ulcers of the mouth, esophagus, stomach, and intestines. Most had severe gum disease and missing teeth. Many had pneumonia or widespread viral or bacterial infections. A large number also suffered from endocrine disorders, including enlargement of and cysts in the thyroid gland. More than half the females examined showed signs of severe breast infections that would have made it difficult, if not impossible, to nurse properly, and nursing mothers had pus mixed with their milk. Some had twisted spines and other skeletal disorders. It was a litany worthy of a cetacean Job.

Shortly after Béland pulled into the vet school with the latest casualty, they began examining the animal—a normal-looking adult male with a healed-over scar on its left side that looked like five circular indentations. The distinctive mark identified him unmistakably as DL-26 (for the belugas' Latin name *Delphinapterus leucas*), one of the previously identified and photographed individuals in the institute's files.

Béland had a sinking feeling when he realized this was Booly, one of the first whales to find a sponsor in the institute's new Adopt-

A-Beluga program, which had been established as a way to raise money for the research program. Just six months earlier, the father of a whale-loving boy in Toronto had sent a check for $5,000; the boy had chosen to christen his adopted beluga Booly.

The autopsy continued into the early hours of the morning and proceeded according to the usual routine—missing teeth, emphysema, extensive stomach ulcers—until they hit the abdominal cavity. Inside, they found two small testicles and normal-looking male plumbing such as the epididymis and vas deferens. But, to their astonishment, Booly had a uterus and ovaries as well—a complete female reproductive tract save for a vagina. All the firsts they had already reported in the St. Lawrence beluga population paled by comparison to this finding. They had discovered the rarest of biological curiosities: a true hermaphrodite. This is a phenomenon seldom seen in wildlife and never before reported in a whale. Even more unusual, Booly had two testes and two ovaries, which previously had been reported scientifically in only two rabbits and a pig. Something had happened to Booly in the womb to derail the normal course of sexual development.

Was Booly an accident of nature or was he a victim of pollution-induced hormone havoc? There is no way to answer this question decades after Booly's birth, but the autopsy report noted: "One cannot rule out that pollutants present in the mother's diet had interfered with hormonal processes" guiding the "normal evolution of the sexual organs of her fetus." Based on the rings of enamel in his teeth, which scientists count like tree rings to estimate a whale's age, the team concluded that Booly was twenty-six years old when he died. He had been born in the early 1960s—a time when the pollution in the St. Lawrence was likely at its peak.

Although pollution levels in the river have dropped markedly since then, the belugas still show high levels of contamination, especially the young. Some of the most contaminated individuals have been under two years of age, and high levels are found even in prematurely born animals who failed to survive, indicating that the contamination had been transferred from the mother across the placenta. After birth, the transfer of contaminants to the offspring continues through the rich, fatty breast milk. While nursing, a mammalian mother (including humans) draws down her fat stores, dumping not only the fat but also the persistent toxic chemicals she has accumulated in her body fat over the years into her milk. In this way, a load of contaminants that it has taken the mother decades to accumulate is passed on to her baby in a very short time. By the time a baby beluga stops nursing at two years of age, it will have acquired a toxic load that, relative to its size, *far exceeds* that of its mother.

Béland and his associates found one young whale that had over five hundred parts per million of PCBs in its body—*ten* times more than the level necessary to qualify as hazardous waste under Canadian law. Ships on the St. Lawrence carrying waste with more than fifty parts per million required a special permit.

Whatever the cause of Booly's unique abnormality, the studies by Béland and his associates suggest there may be widespread hormone disruption among the St. Lawrence belugas, undermining their reproduction and preventing the recovery of the population. The females in the St. Lawrence now have underactive ovaries and a lower rate of pregnancy than Arctic belugas. And based on surveys of younger animals, who have gray skin until they assume the snow white mantle of adulthood at around six years of age, they are producing fewer young than their northern relatives. In the Arctic, young animals make up over forty percent of the population, while the percentage of young animals in the St. Lawrence hovers around thirty, indicating that the reproductive rate is down by twenty-five percent compared to belugas elsewhere.

Such reproductive problems are not surprising. The St. Lawrence belugas carry substantial loads of several synthetic chemicals that are known to disrupt hormones and to interfere with normal reproductive cycles. In a study conducted in the Netherlands, Dutch researchers found impaired reproduction in seals fed contaminated fish from the Baltic, but no problems in a second group fed with cleaner North Atlantic fish. The synthetic chemical contaminants in the Baltic fish are similar to those found in the St. Lawrence fish that the belugas eat.

Besides showing low reproduction, the St. Lawrence beluga population is also experiencing higher than expected mortality among adults in their prime—deaths that the researchers in Quebec think stem partly from the toll that toxic chemicals are taking on their immune systems. There is growing evidence in the scientific literature that both prenatal exposure to hormone-disrupting chemicals and direct adult exposure to toxic compounds can weaken immunity. As with humans suffering from AIDS, the deficient immune systems make the whales more vulnerable to pneumonia, skin diseases, a variety of infections, and cancer. The team is now investigating the beluga immune system and undertaking a comparative study of the St. Lawrence and Arctic populations in an effort to characterize the nature of the immune system damage.

One way or another, Béland says, pollution is killing the whales. But it is not the acute poisoning commonly associated with toxic chemical spills, which causes the quick death of fish and animals caught in the wake. This death is slow, invisible, and indirect.

The discoveries made by Béland, Martineau, and De Guise during their investigation of the St. Lawrence belugas raise broader questions relevant to animal populations everywhere.

As was the case in the St. Lawrence, researchers have commonly blamed the decline and disappearance of wildlife populations on human disruption of their habitat, excessive hunting, fishing, and trapping, or on the introduction of aggressive foreign species that overwhelm native competitors. All these forces are unquestionably at work in the global loss of animal species, but biologists find that they do not explain all the declines.

In light of the growing evidence that many synthetic chemicals disrupt hormones, impair reproduction, interfere with development, and undermine the immune system, we must now ask to what degree contaminants are responsible for dwindling animal populations. Could hormone disruptors account wholly or in part for some losses that have been blamed on classically invoked factors such as habitat loss or overexploitation? Have overexploited species failed to rebound after protection because synthetic chemicals are impairing reproduction?

Asking these questions has already prompted surprising reassessments, even regarding one of the most closely monitored animal populations in the United States—the critically endangered Florida panther. The decline of the big cat, which has come to symbolize the effort to restore the much abused Everglades, had been blamed on reproductive problems caused by inbreeding, human encroachment, road kills, and mercury contamination. Seeking to halt the panther's slide toward extinction, state and federal officials constructed a series of specially designed wildlife underpasses along Alligator Alley, a highway running across the Everglades where a number of panthers have been killed.

The panthers' range in southern Florida, which includes Everglades National Park and the Big Cypress swamp, lies downstream from major agricultural areas and consequently suffers from pesticide and fertilizer pollution. But until quite recently, no one had considered synthetic chemicals a factor in the panthers' plight.

The first clue came in 1989. Prompted by the death of an apparently healthy female in Everglades National Park, federal and state wildlife agencies began a study of the remaining panthers, which number no more than fifty. Wildlife specialists concluded that the female had died from mercury poisoning, which they attributed to the fact that Florida panthers prey heavily on raccoons and are therefore linked through the raccoon to the aquatic food web where mercury and other contaminants accumulate. But the study showed the panthers had a host of other problems as well. These included apparent sterility in some males and females, an extraordinary level of sperm abnormalities, low sperm count, evidence of impaired immune response, and malfunctioning thyroid glands. Thirteen out of seventeen males had undescended testicles, and records on the population showed that the incidence of this problem had increased dramatically in male cubs since 1975. Those investigating the panthers attributed their poor reproduction and related symptoms to lack of genetic diversity resulting from inbreeding in the tiny population.

But as U.S. Fish and Wildlife Service contaminant specialist Charles Facemire became aware of the emerging information on hormone-disrupting synthetic chemicals, he began to question whether bad genes were really the problem. In his research, he had found that the panthers are not particularly inbred compared to other large felines and that their genetic diversity is, in fact, slightly above average. At the same time, he was learning that undescended testicles are a known consequence of prenatal hormone disruption.

If the panthers had suffered hormone disruption in the womb, he learned, it might be evident in the hormone ratios in their blood—specifically in the relative levels of testosterone, the typically male hormone, and estrogen, the typically female hormone. One would expect males to have far higher levels of testosterone, but an analysis of the panthers' blood found ratios that seemed peculiar indeed and suggested that many of the males had been "feminized." Two males had far more estradiol, a form of estrogen, in their blood than testosterone. In several others, estradiol was present at nearly equal levels to testosterone. Although such ratios appear highly abnormal, no definitive conclusion is possible until further work determines normal hormone ratios in other populations of these cats.

The hormone disruption theory took on even more power when Facemire reviewed archived data on contaminants in the animals, for these records showed that the panthers carry high levels of several synthetic chemicals that are known to disrupt hormones. Besides lethal levels of mercury, the fat of the female found dead in 1989 contained 57.6 parts per million of DDE, a breakdown product of the pesticide DDT, as well as 27 parts per million of PCBs, a persistent industrial chemical. At the same time, new findings by Environmental Protection Agency reproductive toxicologist Earl Gray indicate why DDE may be affecting the development of male panther cubs. DDE has long been described as a weak estrogen, but Gray's studies have demonstrated that it is also a potent *blocker* of *male hormones*. If something blocks testosterone messages while a male is developing in the womb, Facemire notes, undescended testicles might be one of the consequences, for testosterone cues their descent from the abdomen into the scrotum late in gestation.

Despite the uncertainties about the cause of the panthers' predicament, wildlife managers are nevertheless proceeding with an expensive translocation program that will bring panthers from healthier out-of-state populations to Florida. The hope is that these imported cats will mate with local panthers and reduce the reproductive problems that some still attribute to inbreeding.

It may take several years to settle the question of what has brought the Florida panther to the brink of extinction, but with this preliminary work, hormone disruption caused by contaminants has suddenly emerged as the most compelling theory to explain their reproductive problems. In a number of other species, however, scientists have already established persuasive evidence linking hormone-disrupting chemicals to reproductive failures, and in many of these cases, the losses had been blamed initially on other factors.

One of the best examples is the work done by biologists from the University of Florida, the U.S. Fish and Wildlife Service, and the Florida Game and Freshwater Fish Commission on the alligator population in Lake Apopka. The 12,500-hectare lake, the fourth-largest freshwater body in the state, is located far north of the Everglades, not far from Orlando and Disney World. Although University of Florida reproductive biologist Lou Guillette knew the reproductive failure had to be related to the Tower Chemical Company spill on the shore of Lake Apopka more than a decade earlier, the symptoms in the wildlife began to make sense only when he discovered that synthetic chemicals could act like hormones.

That revelation came, Guillette recalls, on a lazy Friday afternoon in the spring of 1992, when his longtime mentor, Howard Bern, a comparative endocrinologist and professor emeritus from the University of California at Berkeley, gave an informal lecture during a visit to the Gainesville campus. In that talk, Bern spoke about his pioneering work on DES and about the discovery that similar kinds of developmental disruption have occurred in birds exposed to synthetic chemicals. Bern had himself just learned of the striking and troubling parallels when he had attended an interdisciplinary conference the previous July at the Wingspread Conference Center in Racine, Wisconsin. To Guillette's amazement, the problems Bern described were the same as those found in the alligators he had been studying.

Because of the prevailing assumptions about contaminants, Guillette and his colleagues had expected the chemical spill to cause cancer or death. Bern's talk gave them a whole new way to think about how contaminants might affect alligators and other organisms and a whole new set of questions. "We knew it was contamination," he says, "but we didn't realize it was a hormone effect."

With this insight, Guillette could now imagine how the unusual problems seen in the alligators might be connected to the 1980 chemical spill, an event that placed the area on the federal Superfund list of the nation's most serious hazardous waste sites. The chemical released in the spill was dicofol, a close pesticide relative of DDT and a compound that also interferes with hormones. Since then, Guillette, along with Timothy Gross of the University of Florida, Franklin Percival of the U.S. Fish and Wildlife Service, and Allan Woodward of the Florida Game and Freshwater Fish Commission, have been gathering the data that support such a connection.

After the issue first broke into the news in early 1994, a parade of journalists began trooping down to Lake Apopka to record the plight of its alligators and photograph their tiny members, which are only one-third to one half normal size. The problem is more severe in those living near the site of the chemical spill than in those living in the northern part of the lake five to ten miles away, Guillette and his colleagues have found. But regardless of where they live on the lake, the Apopka males as a whole have smaller penises than males hatched on a relatively clean lake. More recent studies also show that this problem extends beyond Lake Apopka, although the symptoms are less severe in lakes with no history of industrial chemical spills. Researchers are now exploring the possibility that agricultural contamination is responsible.

Even if these males had normal-size penises, the Apopka alligators would still be having reproductive difficulties because both sexes suffer from profound but less visible disruption in their internal reproductive organs. The ovaries in the female alligators display abnormalities in their eggs and in the follicles, where the eggs mature before ovulation, that are remarkably similar to abnormalities reported in humans and lab animals exposed to DES early in development. The males' testicles showed structural defects as well.

The Apopka alligators also have skewed hormone ratios, with males showing a profile that looks typical of a normal female. These males have elevated levels of estrogen and greatly reduced levels of testosterone in their blood—only one-fourth the level found in males from the relatively uncontaminated Lake Woodruff. The females from Apopka also have elevated estrogen levels and an estrogen-to-testosterone ratio that is twice as high as normal. These significantly aberrant hormone levels in males and females suggest their sex organs may function poorly or not at all.

To Guillette's surprise, the contamination on Apopka has also taken a devastating toll on its red-eared turtle population. Unlike the carnivorous alligators, which are top predators and therefore exposed to higher levels of contaminants that have become more concentrated in the journey up the food web, the red-eared turtles also eat plants—a dietary habit that should expose them to less pollution. Nevertheless, they, too, are having reproductive difficulties. In their case, however, the problem is not unhatched eggs but an absence of males. Researchers find females in the lake and many turtles that are neither male nor female. Because of hormone disruption during sexual development, the animals that would have become males end up stranded in the gender-bending state called intersex. There are few normal males to be found.

In turtles, sex is determined by temperature rather than by a gene, leading some to suggest that this intersex condition could be a natural abnormality caused by temperature fluctuations during the incubation of the eggs. But subsequent research by David Crewes of the University of Texas at Austin has shown that such aberrations do not show up even when turtle eggs are incubated at a borderline temperature for producing one sex or the other. Something more than temperature fluctuation is needed to produce sexually confused animals. In laboratory experiments, researchers have been able to produce intersex animals by incubating turtle eggs at a male temperature and exposing them at the same time to estrogen or estrogenic synthetic chemicals such as some of the PCBs.

While the effects of the contaminants on the alligators and turtles are obvious, the team has not yet determined which chemical or chemicals are responsible. The leading suspect is the DDT breakdown product DDE, the contaminant found in the highest concentrations in the alligator eggs and one that could originate from the spilled dicofol, but other endocrine-disrupting contaminants are present as well, including dieldrin and chlordane. In tests in which researchers painted alligator eggs with DDE, as little as one part per million was enough to produce greater than expected rates of ambiguous development in the sex organs—aberrations resembling those reported in contaminated birds by University of California researcher Michael Fry. Chemical analyses of Apopka alligator hatchlings have found that they contain four to five parts per million of DDE.

Given these severe reproductive problems, one would expect to find few if any alligators, but that is far from the case. Researchers believe the immigration of healthy alligators from cleaner lakes to Apopka has prevented their disappearance. It is not unusual for Florida alligators to move among lakes looking perhaps for better habitat and for better hunting, so a lake such as Apopka—with prime habitat and a depleted native alligator population—would be bound to attract migrants. This constant replenishment from outside has totally masked the lake's severe problems. Indeed, Apopka's dire condition might have gone undiscovered if wildlife officials hadn't become interested in supplying wild alligator eggs for commercial alligator ranches. Faced with the question of how many eggs could be collected from the wild without jeopardizing the once endangered species, Guillette's colleagues surveyed several Florida lakes including Apopka. In this way, they found that most of the eggs in Apopka nests were not hatching.

Lake Apopka vividly illustrates how appearances can be totally at odds with reality. The lake appears to be a healthy, relatively unspoiled place, and the surrounding swamps seem to be rich in wildlife, including turtles and alligators. By ordinary water quality measures, the 1980 spill is ancient history and the lake is again clean. But even years later the poisons from the accident have not truly disappeared. Though absent from the water, they are still circulating in Apopka's food web and causing havoc. Only with closer scrutiny does the profound disruption of its wildlife become evident.

Immigration of this kind—where animals born in relatively clean areas constantly move into contaminated habitats and settle—is also hiding continuing problems in other wildlife species in the

United States, including the bald eagle. After the federal government restricted use of DDT and dieldrin in the early 1970s, eggshell
thinning diminished, and by the 1980s, the bird began making a remarkable recovery across the country. By 1994, the U.S. Fish and
Wildlife Service proposed removing the bird from the federal endangered species list. Nevertheless, some of the nation's bald eagle populations remain troubled.

In the Great Lakes, bald eagle numbers climbed from 26 to
134 pairs between 1977 and 1993, but this recovery may be more apparent than real. U.S. Fish and Wildlife Service biologists believe
that the growth of the Great Lakes population depends largely on
immigration of eagles hatched in cleaner areas. These recruits from
inland areas of Michigan, Minnesota, and Wisconsin often breed
well at first but become less successful as contaminants from the
lakes accumulate in their bodies and impair their fertility. Studies
have found that the birds take in the contaminants through their
prey and that the higher the levels of DDE and PCBs, the lower
their breeding success.

An eagle's diet determines how quickly it acquires sufficient
levels of contaminants to impair reproduction. Although Lake Superior is less polluted than the other Great Lakes, its eagles accumulate
an appreciable amount of contamination because of their eating
habits. Rather than preying directly on fish, Lake Superior eagles frequently prey on fish-eating birds such as gulls—one step higher in
the food web. Because of this taste for gulls, they accumulate concentrations of contaminants that are twenty times greater than they
would have if they had eaten only fish.

Poor reproduction among bald eagles continues to be a problem in several other areas of the country as well, affecting populations in the lower Columbia River in the Pacific Northwest, in an
area near Yellowstone National Park, and in Maine's coastal areas,
where eagle populations may prey more heavily upon birds.

Impaired reproduction in adults was just one of the problems
that emerged in the Great Lakes. In eagles and other fish-eating
birds, field biologists began to see severe birth defects such as missing eyes, clubbed feet, and crossed bills and a bizarre wasting syndrome that would suddenly strike an apparently healthy chick and

cause it to wither and die. Here again, low genetic diversity was initially invoked to explain the problems. Those offering this explanation reasoned that DDT had almost totally wiped out some species in the region, so the animals in the rebounding populations were all likely to be descendants of the small number of animals that had survived the crash and hence inbred.

Based on environmental detective work and sophisticated toxicology, scientists now have evidence that this, too, is a case of contamination rather than inbreeding. For several decades, researchers believe, DDT-induced eggshell thinning masked the effects of other contaminants by killing the embryo. With reduced use of DDT, eggshell thinning abated and chicks began to survive, allowing other physical and behavioral abnormalities to emerge. Some of these problems have been linked to dioxins along with furans, and certain PCBs that act through the same mechanism As described in Chapter 7, all these chemicals behave in a dioxinlike manner and bind to an orphan receptor, whose normal function in the body remains unknown.

In 1993, the bald eagles nesting along the Great Lakes produced four deformed chicks—three with crossed bills and one with malformed feet—but such problems are not limited to the Great Lakes. In the summer of 1994, the U.S. Fish and Wildlife Service began investigating a cluster of similar deformities among birds in Oregon, where nine birds in the Rogue Valley, including red-tailed hawks, kestrels, an osprey, a Brewer's blackbird, and a robin, were found with crossed bills, missing eyes, or both. Wildlife specialists say that a crossed bill is a developmental deformity analogous to a cleft palate in mammals.

As many of these cases indicate, adult animals can tolerate levels of pollution that devastate their offspring. But in the most sensitive species, such as the mink, pollution levels in the Great Lakes appear to be still too great even for adult survival. Although records are sparse, it appears that mink began disappearing from the shoreline of the Great Lakes in the mid-1950s, and even with the restrictions on DDT, PCBs, and other persistent chemicals they have not yet returned. Some have blamed their disappearance on the destruction of habitat and human disturbance, but wildlife specialists doubt this is an adequate explanation, in part because of the abundance of muskrats at the water's edge. Mink and muskrats prefer similar habitats, so conditions hospitable to one should be suitable to the other. The reason muskrats thrive and mink are absent is likely related to their diets. Muskrats eat plants; mink, which are carnivores, eat higher on the food web, and those inhabiting the shoreline consume a good deal of contaminated fish.

Following the deaths and reproductive failures in the 1960s among ranch mink fed on Great Lakes fish, Michigan State University biologists Richard Aulerich and Robert Ringer conducted a series of studies attempting to discover what was killing the animals—work that also shed light on the fate of their wild cousins. The pair discovered that DDT and pesticides were not the problem; the mink were dying because they are highly sensitive to PCBs. Females failed to reproduce when they were fed food containing from 0.3 to 5 parts per million of PCBs—levels found in Great Lakes fish today and in the fat of human breast milk. Adult animals died at doses of 3.6 to 20 parts per million.

In Britain and Europe during the 1950s, a parallel decline occurred among otters, a winsome relative of the mink that feeds primarily on fish, and they, too, are still absent from many areas where they were widespread four decades ago. Although otters face a constant threat from habitat destruction throughout much of Europe, two British specialists, Sheila Macdonald and Chris Mason, found strong evidence that local and wind-borne pollution may be equally important factors in the disappearance of the otter from many areas such as eastern England where the native population appears to be extinct.

In analyzing field surveys done in Great Britain in the 1970s and in a number of European countries in the 1980s, Macdonald and Mason discovered that otter populations are thriving only on the Atlantic seaboards in countries such as Norway, Scotland, and Ireland; in southwest France and western Spain, where the prevailing winds blow off the ocean; and in southeast Europe, especially Greece. But otters are in a perilous state in highly industrialized countries and in regions downwind from major industrial areas. The otters, like their relative the mink, appear to be sensitive to PCBs, and all evi-

dence indicates that PCBs, acting alone or in concert with other contaminants such as mercury, had a significant role in their disappearance. Studies on the threatened otter populations in Oregon and
Sweden have found a similar association with PCBs.

Some fish species also appear to be extremely sensitive to certain synthetic chemicals, though fisheries specialists have been slow
to recognize their role in collapsing fisheries. The lake trout was extinct in Lake Ontario by the early 1950s and crashed during this
same period elsewhere in the other Great Lakes as well. Today, this
native top predator reproduces naturally only in Lake Superior and
parts of Lake Huron, so the presence of the species is maintained
largely through artificial stocking programs.

The demise of the native lake trout has been attributed to a
combination of overfishing, habitat destruction, and predation by
the sea lamprey, an exotic parasite that attaches itself to other fish
and sucks their body fluids, but recent studies are now challenging
this explanation. Although the contaminant levels found in the lakes
do not kill adult fish, experiments have shown that trout eggs are extremely sensitive to dioxin and other chemicals, such as some PCBs,
that share the same toxic mechanism. They appear to be the most
sensitive species ever tested. When exposed to doses of as little as
forty parts per trillion of dioxin or 2,3,7,8-TCDD (or its toxic equivalent), trout eggs or newly hatched fish begin to show significant mortality. By one hundred parts per trillion, all the eggs die. Based on
core samples taken of the lake sediment, EPA toxicologists are reconstructing the history of dioxin exposure in the lake using com
puter models. This work indicates that dioxin reached sufficient
levels by the 1940s in Lake Ontario to begin killing trout eggs and seriously undermining their reproduction.

Growing evidence that dioxin and dioxinlike PCBs are in part
or wholly responsible for the loss of the lake trout raises additional
questions about whether contamination has claimed other species
as well. The lake trout fed on the deep-water sculpin, a species that
was never commercially harvested, but it, too, has vanished. DDT,
dioxin, and PCBs bind most effectively to the smallest organic particles, which are constantly sorted and resorted by currents within the
lakes until they are finally deposited in the muds of the deepest wa-

ter, where the deep-water sculpin lived. As a result, that species was exposed to much higher levels than other fish.

Egg mortality is not the only factor undermining the survival among Great Lakes fish. In the 1960s, fish and game agencies introduced several species of salmon from the Pacific Northwest, which are now stocked annually from fish hatcheries and provide the basis for a multibillion-dollar sport fishing industry in the region. Today, all the salmon in the Great Lakes have severely enlarged thyroid glands, a symptom of inadequate levels of thyroid hormone, a chemical messenger that plays a role in reproduction and in the development of healthy offspring. Salmon eggs normally contain high levels of thyroid hormone, but studies have found that Great Lakes salmon eggs have lower thyroid hormone levels than eggs from Pacific salmon in less contaminated areas. Although iodine deficiency can cause such symptoms, researchers have ruled it out as the cause. Researchers have found that Great Lakes herring gulls, which prey on fish, also suffer from thyroid enlargement. All evidence indicates that these thyroid problems stem from contamination, and studies have shown that many of the chemicals found in the lakes can block the action of thyroid hormones, but the specific chemicals responsible have not been identified.

The salmon in Lake Erie also show a variety of reproductive and developmental problems, most notably precocious sexual development and a loss of typically male secondary sexual characteristics such as heavy protruding jaws and red coloration on the flanks. Although secondary and sexual characteristics have been known to diminish in hatchery fish because of the absence of natural selection, John Leatherland, a fish specialist at the University of Guelph in Ontario, Canada, does not believe that the evolutionary phenomenon known as "genetic drift" is sufficient to explain the significant loss of visible sex differences in these fish, as some have suggested. Genetic drift is an evolutionary process that occurs in very small populations where chance begins to play a much larger part in determining which genetic characteristics appear in the next generation. In this case, it becomes an alternative to natural selection, the evolutionary process at work in large populations. But efforts to explore whether endocrine-disrupting chemicals are responsible for the changes seen in the salmon have been stymied by ignorance about basic physiology in this family of fish and about the processes that determine the expression of secondary sexual characteristics.

Many animals show reproductive problems that have been linked to contaminants, but marine mammals such as whales, dolphins, seals, and polar bears may face the greatest jeopardy, particularly over the long term. As illustrated by the journey of the PCB molecule in Chapter 6, persistent chemicals accumulate and concentrate greatly in the marine food web, exposing the long-lived predators to high levels of contamination. These marine animals are especially vulnerable to persistent chemicals because they carry heavy layers of fat that serve both as insulation against cold waters and reserve fuel for times when food is in short supply. With time, the vast quantities of persistent chemicals released over the past half century on land, such as PCBs, will gradually make their way into the oceans and increase already threatening levels of contamination.

In Europe and Scandinavia, researchers have done a series of studies, beginning in the early 1970s, on declining populations of harbor, ringed, and gray seals in the Baltic and on harbor seals in the Wadden Zee, a large extension of the North Sea off the coasts of Holland, Germany, and Denmark. Like the belugas in the St. Lawrence, the harbor seal population off the Dutch coast dropped from about 1,540 to 550 between 1965 and 1975 even though hunting had stopped.

Taken together, these studies show significantly decreased reproduction among seals living in contaminated areas, which is strongly correlated with their levels of PCBs. In the Baltic, high body burdens of PCBs are linked to a higher incidence of abnormalities in female ringed seals, including tumors and other obstructions in the uterus that can result in the death of their offspring. Studies on Dutch harbor seals have also found a strong correlation between the decrease in reproductive success and high levels of PCBs in their tissue. In follow-up experimental studies mentioned earlier in this chapter, Dutch researcher Peter J. H. Reijnders fed two groups of female common seals fish from two sources—the contaminated Dutch

Wadden Zee and the cleaner northeast Atlantic. In breeding season, the females from both groups consorted with three males who had been eating uncontaminated Atlantic fish. The difference in the pregnancy rate between the two groups was remarkable. Ten out of the twelve females eating cleaner Atlantic fish became pregnant, but only four in the second group that had eaten Wadden Zee fish.

Then in the late '80s, a series of dramatic marine epidemics began to kill off thousands of seals, dolphins, and porpoises, hitting populations in the Baltic and North Seas, the Mediterranean, the Gulf of Mexico, the North Atlantic, the eastern coast of Australia, and even seals in Lake Baikal in Siberia. The losses devastated several populations:

- In 1987, a distemper virus claimed an estimated ten thousand seals in Lake Baikal.
- That same year the Atlantic coast from New Jersey to Florida witnessed a die-off of bottlenose dolphins that extended into 1988 and claimed more than 700 victims—a loss of more than half of that migratory coastal population.
- In 1988, twenty thousand harbor seals, up to 60 percent of some local populations, perished in a matter of months in the North Sea.
- Between 1990 and 1993, more than a thousand striped dolphins washed up on the shores of the Mediterranean.

Are all these die-offs a bizarre coincidence or evidence of some serious and pervasive problem among marine mammals?

In three cases, the animals had succumbed to infections caused by distemperlike viruses. In others, bacteria or fungi were implicated as the "causative agents." But even if the immediate cause of death seemed clear, this does not answer the deeper question of why so many animals in different places have become so vulnerable. The autopsies on the victims did, however, yield some suggestive clues. The dead animals showed weakened immune systems, and those examined carried high levels of contaminants such as PCBs, synthetic chemicals that have been shown to suppress immune response in numerous animal studies.

Distemper virus itself certainly contributed to immune system suppression among the victims of these outbreaks of virus. But two recent studies have suggested that contaminant-induced immune suppression may also have been involved.

Virologist Albert Osterhaus and a team of Dutch researchers at the National Institute of Public Health and Environmental Protection undertook another feeding study with two groups of seals to see whether eating contaminated fish had any effect on the animals' immune response. As in the earlier study looking at reproduction, half of the twenty-two seals in the study ate relatively clean North Atlantic fish; the other half ate fish from the heavily polluted Baltic, which contained roughly ten times more dioxinlike organochlorines and similar-acting compounds than the Atlantic herring. Osterhaus has noted in press interviews that all the fish fed to both groups had come from commercial catches destined for human consumption.

The seals that dined on Baltic herring for nearly two years quickly showed several signs of depressed immune function and weakened ability to fight off viral infections. The action of their natural killer cells, which act as a first line of defense against viral infections, was twenty to fifty percent below normal and the response of T cells, which direct the immune defense, dropped by twenty-five to sixty percent. T lymphocytes play a crucial role in clearing viral infections, especially in the case of distemperlike viruses.

A second study done in the United States by immunologist Garet Lahvis looked at the relationship between immune response and pollutant levels in relatively healthy dolphins, whose blood was taken after they were encircled by nets in shallow water off the coast of Florida. Using lymphocyte assays similar to those used to detect early signs of immune impairment in humans infected with the AIDS virus, Lahvis found evidence that immune response in dolphins dropped as levels of PCBs and DDT increased in their blood. This provides added evidence that highly contaminated animals were already likely to be suffering from weakened immune systems before they were attacked by the distemperlike viruses implicated in some of the marine epidemics.

Like the reproductive system, the immune system is especially vulnerable to damage from hormone-disrupting chemicals during prenatal development. As we've seen, animal studies and evidence from DES-exposed humans show that such exposure can alter the development of the immune system and have a lifelong impact. The growing evidence of the links between synthetic chemicals and immune system impairment has particularly serious implications for marine mammals. These species carry high levels of contamination, making it likely their offspring enter the world with already impaired immune systems. The chronic exposure to toxic chemicals afterward during a lifetime in polluted waters simply adds insult to injury. Because of their prenatal exposure, the offspring may be less able to fight off disease than their parents, whose exposure to chemicals was limited to their adulthood.

If synthetic chemicals have weakened marine mammal immune systems and contributed to the rash of epidemics, one might wonder why the outbreaks didn't occur earlier. Part of the reason may be that these are long-lived species, so it would take some time for this second-generation double whammy to emerge. There is also a lag time between the release of contaminants on land and their accumulation in the sea. These two factors may explain in part why these major marine epidemics did not show up until recently.

Even less is known about the cause of the dramatic and mysterious frog declines that have been reported in many parts of the world. The loss of frogs in urban areas of the United States where development has gobbled up wetlands seems no mystery, but why are frogs vanishing in undisturbed forest in Costa Rica and in remote areas of Australia? One of the world's leading amphibian experts, Robert Stebbins, believes that hormone-disrupting chemicals are likely suspects in declines that do not have obvious causes such as habitat disruption or drought. Speaking at an international gathering of herpetologists in Australia in December 1993, Stebbins, professor emeritus of zoology at the University of California at Berkeley, urged his colleagues to give the possible role of hormone-disrupting chemicals high priority in their search for a cause.

During his research on amphibians, Stebbins has reviewed the reports of frog declines and found a synchronous pattern in many that suggests a widespread cause such as wind-borne contaminants.

Based on his review, it appears that many populations declined rapidly or vanished altogether between the mid-1970s and early 1980s. Populations living at high altitudes have been particularly hard hit, leading some to suggest that the thinning ozone layer might be playing a role by exposing them to increased levels of damaging ultraviolet-B radiation. While depleted ozone may be undermining some frog populations, Stebbins has not found this theory adequate to explain losses such as the golden toad of Costa Rica, a species that lives deep in the rain forest and is well protected from UVB radiation by the trees.

High-altitude areas are not only more vulnerable to ozone loss, they are also more vulnerable to pollution that arrives on the winds. As our hypothetical journey of the PCB molecule illustrated, many persistent synthetic chemicals evaporate and travel around until they hit a cooler spot, such as a mountain, where they recondense and come to rest. As scientists have studied the role of long-distance pollution in acid rain, they have found all kinds of contaminants on seemingly remote mountainsides.

Stebbins sees other symptoms that seem to implicate chemical contamination as well. Like the dolphins, some of the disappearing frogs seem to be showing signs of weakened immune systems, as indicated by outbreaks of "red leg." This sometimes lethal infection, which causes inflammation on the underside of the legs, is caused by a common bacterium that is found in freshwater around the world.

"If a population is healthy," Stebbins says, "the frogs can handle this bacterium quite well. But in a lot of places where there are declines, animals are showing up with red leg." Because of the apparent relationship to a weakened immune system, Stebbins has described the phenomenon as "AIDS-like."

Finally, Stebbins notes, the frog's unique physiology and life history make it highly vulnerable to the assault of hormone-disrupting chemicals. Endowed with permeable skin that can breathe and take in water, frogs absorb the chemicals they encounter more readily than most other animals. They also experience a dramatic passage called metamorphosis that transforms them from creatures that breathe in water to creatures that breathe in air. As tadpoles become frogs, a profound reorganization of their structure and physiology takes place—a process driven by hormones and therefore vulnerable to synthetic chemicals that disrupt hormone messages. By their nature, frogs are prime candidates for hormone havoc.

And many other creatures may be as well. Scientists are just beginning to explore the possible impacts of pesticides and other synthetic chemicals on the migration of birds, a complex and awe-inspiring phenomenon that is still not fully understood. Some small birds fly as high as airplanes; some make ninety-hour nonstop flights from Massachusetts to South America. Two-thirds of the bird species in North America are migratory, and many of the long-distance migrants have shown alarming declines.

In the mid-1980s, Pete Myers, then a young scientist studying shorebirds, began to wonder about the possible impact of pesticides as he contemplated the plunging numbers of shorebirds, a family that includes such species as sandpipers and plovers. Shorebirds perform some of the greatest migratory feats in the avian world, with some species traveling well over fifteen thousand miles in a single year, from breeding sites in the Arctic to wintering grounds in far southern South America. Data from the Manomet Bird Observatory in Plymouth, Massachusetts, suggested that the population of sanderlings moving south along the East Coast of the United States in autumn had declined eighty percent over a fifteen-year period. Much of this population, Myers had recently discovered in his research, wintered on the west coast of South America in Peru and Chile, where he was working.

What was happening to them and why? The birds cover so much ground in the course of a year hopscotching their way back and forth across the equator, depending not only on breeding and wintering grounds but on important stopover sites along the way where they feed and build up energy for the next leg of their journey. The destruction of just one of these key feeding areas along the way might derail their marathon migrations and doom the birds.

The sanderling decline did not appear to be caused by a loss in winter habitat, for there was an abundance of sandy beaches where the birds could feed and rest. Perhaps something had gone wrong in the breeding grounds in the Arctic in Alaska and Canada or in one of the key feeding sites such as Delaware Bay or the coast of Massachusetts. Myers and his colleagues began to tackle these habitat questions, working to identify the precise migratory routes the birds were traveling and the critical way stations in their journey.

But Myers suspected that there might be another factor to consider as well. Every day he watched the birds congregate to feed and bathe at the mouths of the rivers and streams that emptied intensely cultivated river valleys in the otherwise harsh Peruvian desert. Virtually every rivulet and stream, no matter how small, reeked of pesticides used in the cotton and rice fields packed tightly into verdant river valleys. As the birds put on weight in preparation for heading north again, it appeared likely they were taking on a load of contaminants carried in their food. Where did those pesticides go when the birds suddenly burned off that fat during a marathon flight from South America to Cape Cod? As the stored fat is converted to energy for their flight, the contaminants would be liberated into the blood and most likely move to either the sex organs or the brain, both of which contain major fat deposits. If that was happening, the pesticides might be interfering with migration, disrupting reproduction, or even killing the birds. Would a contaminated brain steer a bird off course? If they reached their breeding grounds, would they be able to reproduce? And what would happen to the offspring of young born to contaminated adults? Would their behavior and ability to orient in migration be impaired? Science was not prepared to answer, and even today, it is unknown whether pesticides played a role in the decline of the sanderlings and other migratory species over the past four decades.

With the growing amount of evidence and theories that link wildlife problems to hormone disruption, there is now good reason to regard endocrine-disrupting chemicals as a major long-term threat to the world's biodiversity and perhaps an immediate threat to certain endangered species, such as the St. Lawrence belugas and the Florida panthers. In searching for the causes of loss, scientists must look for functional changes, such as impaired reproduction and altered behavior, along with more evident disruptions, such as vanishing habitat and changing climate. As many of the cases discussed in this chapter demonstrate, it is important to look beyond appearances. Animals that appear normal and healthy may, in fact, show skewed hormone ratios, scrambled sex organs, or physical changes in their brains when one takes a closer look. Diminished and impaired by invisible damage, such animals lose the edge honed by millions of years of natural selection. They may lose their ability to withstand otherwise tolerable stresses or to rebound after natural disasters. For no apparent reason, they may suddenly disappear or slowly, imperceptibly slip into extinction.

10

ALTERED DESTINIES

"OUR FATE IS CONNECTED WITH THE ANIMALS," RACHEL CARSON WROTE more than three decades ago in *Silent Spring*, a now classic indictment of synthetic pesticides and human hubris that helped launch the modern environmental movement. This has long been a guiding belief among environmentalists, wildlife biologists, and others who recognize two fundamental realities—our shared evolutionary inheritance and our shared environment. What is happening to the animals in Florida, English rivers, the Baltic, the high Arctic, the Great Lakes, and Lake Baikal in Siberia has immediate relevance to humans. The damage seen in lab animals and in wildlife has ominously foreshadowed symptoms that appear to be increasing in the human population.

As noted in earlier chapters, basic physiological processes such as those governed by the endocrine system have persisted relatively unchanged through hundreds of millions of years of evolution. Evolutionary narratives tend to highlight the innovations of natural selection, ignoring the stubborn conservative streak that has marked the history of life on Earth. At the same time that evolution experimented greatly with form, shaping the vessels in various and wondrous ways, it has strayed surprisingly little from an ancient recipe for life's biochemical brew. In examining our place in the evolutionary lineage, humans tend to focus inordinately on those characteristics that make us unique. But these differences are small, indeed, when compared to how much we share not only with other primates such as chimpanzees and gorillas, but with mice, alligators, turtles, and other vertebrates. Though turtles and humans bear little physical similarity, our kinship is unmistakable. The estrogen circulating in the painted turtle seen basking on logs during lazy summer afternoons is exactly the same as the estrogen rushing through the human bloodstream.

Humans and animals share a common environment as well as a common evolutionary legacy. Living in a man-made landscape, we easily forget that our well-being is rooted in natural systems. Yet all human enterprise rests on the foundation of natural systems that provide a myriad of invisible life-support services. Our connections to these natural systems may be less direct and obvious than those of an eagle or an otter, but we are no less deeply implicated in life's web. No one has stated this fundamental ecological principle more simply than the early twentieth-century American environmental philosopher, John Muir. "When we try to pick out anything by itself, we find that it is bound by a thousand invisible cords . . . to everything in the universe."

Our regrettable experience with persistent chemicals over the past half century has demonstrated the reality of this deep and complex interconnection. Whether we live in Tokyo, New York, or a remote Inuit village in the Arctic thousands of miles from farm fields or sources of industrial pollution, all of us have accumulated a store of persistent synthetic chemicals in our body fat. Through this web of inescapable connection, these chemicals have found their way to each and every one of us just as they have found their way to the birds, seals, alligators, panthers, whales, and polar bears. With this shared biology and shared contamination, there is little reason to expect that humans will in the long term have a separate fate.

Yet, some skeptics question whether animal studies provide a useful tool for forecasting threats to humans. The refrain that "mice are not little people" has been heard frequently in the ongoing debate about whether animal testing accurately predicts whether a chemical poses a cancer risk to humans. Critics have also attacked testing procedures for using unrealistically high doses, arguing, for example, that mice tested to discover if DDT caused cancer were fed more than eight hundred times the average amount humans would take in eating a typical diet.

Whatever the merits of these criticisms regarding cancer testing, they have little relevance to the use of animals to predict the effects of hormone-disrupting chemicals. Because scientists have only an incomplete understanding of the basic mechanisms that induce cancer, extrapolating from one species to another has admitted uncertainties. In contrast, scientists have a good grasp of the mechanisms and actions of hormones. They understand how chemical messages are sent and received and how some synthetic chemicals disrupt this communication. They know that hormones guide development in basically the same way in all mammals, and if there was any doubt, the DES experience has verified the similarity of disruption across many species, including humans. Time after time, the abnormalities first seen in laboratory experiments with DES later showed up in the children of women who had taken this drug during pregnancy.

The relevance of the DES experience to the threat from environmental endocrine disruptors has also been questioned because of the very high doses given to pregnant women and to laboratory animals in experiments. While most of the early experiments did indeed use high doses, recent studies using much lower doses have produced no less alarming results. In fact, in some cases a high dose may paradoxically cause *less* damage than a lower dose. In exploring the effects of much lower doses of DES, Fred vom Saal has found that the response increases with dose for a time and then, with even higher doses, begins to diminish.

Vom Saal's dose response curve looks like an inverted U. Its shape is profoundly important to the interaction between the endocrine system and synthetic contaminants. Neither linear nor always moving in the same direction, the inverted U seems characteristic of hormone systems and it means that they do not conform to the assumptions that underlie classical toxicology—that a biological response always increases with dose. It means that testing with very high doses will miss some effects that would show up if the animals were given lower doses. The inverted U is another example of how the action of endocrine disruptors challenges prevailing notions about toxic chemicals. Extrapolation from high-dose tests to lower doses may in some cases seriously underestimate risks rather than exaggerate them.

Because the endocrine disruption question has surfaced so recently, the scientific case on the extent of the threat is still far from complete. Nevertheless, if one looks broadly at a wide array of existing studies from various branches of science and medicine, the weight of the evidence indicates that humans are in jeopardy and are perhaps already affected in major ways. Taken together, the pieces of this scientific patchwork quilt have, despite admitted gaps, a cumulative power that is compelling and urgent.

This was the lesson from the historic meeting on endocrine disruption that took place in July 1991 at the Wingspread Conference Center in Racine, Wisconsin. Over the years, dozens of scientists have explored isolated pieces of the puzzle of hormone disruption, but the larger picture did not emerge until Theo Colborn and Pete Myers finally brought twenty-one of the key researchers together. At this unique gathering, specialists from diverse disciplines ranging from anthropology to zoology shared what they knew about the role of hormones in normal development and about the devastating impacts of hormone-disrupting chemicals on wildlife, laboratory animals, and humans. For the first time, Ana Soto, Frederick vom Saal, Michael Fry, Howard Bern, John McLachlan, Earl Gray, Richard Peterson, Peter Reijnders, Pat Whitten, Melissa Hines, and others explored the exciting connections between their work and the ominous implications that arose from this exercise. As the evidence was laid out, the parallels proved remarkable and deeply disturbing. The conclusion seemed inescapable: the hormone disruptors threatening the survival of animal populations are also jeopardizing the human future.

At the end of the session, the scientists issued the Wingspread Statement, an urgent warning that humans in many parts of the world are being exposed to chemicals that have disrupted development in wildlife and laboratory animals, and that unless these chemicals are controlled, we face the danger of widespread disruption in human embryonic development and the prospect of damage that will last a lifetime.

The pressing question is whether humans are already suffering damage from half a century of exposure to endocrine-disrupting synthetic chemicals. Have these man-made chemicals already altered individual destinies by scrambling the chemical messages that guide development?

Many of those familiar with the scientific case believe the answer is yes. Given human exposure to dioxinlike chemicals, for example, it is probable that some humans, especially the most sensitive individuals, are suffering some effects. But whether hormone-disrupting chemicals are now having a broad impact across the human population is difficult to assess and even harder to prove. This is inescapable in light of the nature of the contamination, the transgenerational effects, the often long lag time before damage becomes evident, and the invisible nature of much of this damage. Those trying to document whether perceived increases in specific problems reflect genuine trends in human health find themselves thwarted by a dearth of reliable medical data. Few disease registries exist for anything except cancers. A number of pediatricians from various parts of the United States have expressed their concern about an increasing frequency of genital abnormalities in children such as undescended testicles, extremely small penises, and hypospadias, a defect in which the urethra that carries urine does not extend to the end of the penis, but it is virtually impossible to document these anecdotal reports. Unfortunately, the problems caused by endocrine disruption may have to reach crisis proportion before we have a clear sign that something serious is happening.

In the face of these difficulties, the animal studies provide a touchstone for identifying and investigating what might be happening in humans. They can alert us to the probable kinds of disruptions and help focus research efforts. They can also provide early warnings about the hazards of current levels of contamination. Because of the diversity of life, some animals are bound to be more readily exposed to contaminants than humans. Transgenerational effects, such as changes in behavior and diminished fertility, are also likely to show up faster in wildlife because most animals mature and reproduce more quickly than humans. Experimental work with animals adds another equally invaluable dimension. As the history of DES demonstrates, laboratory experiments with rats and mice accurately forecasted damage that later showed up in humans. The tragedy is that we ignored the warnings.

Based on the warnings from wildlife and lab animals, what kinds of problems should we expect? Earlier chapters explored how hormonally active synthetic chemicals can damage the reproductive system, alter the nervous system and brain, and impair the immune system. Animals contaminated by these chemicals show various behavioral effects, including aberrant mating behavior and increased parental neglect of nests. Synthetic chemicals can derail the normal expression of sexual characteristics of animals, in some cases masculinizing females and feminizing males. Some animal studies indicate that exposure to hormonally active chemicals prenatally or in adulthood increases vulnerability to hormone-responsive cancers, such as malignancies in the breast, prostate, ovary, and uterus.

Is there evidence of such problems in humans? Are these problems increasing? In some instances, it appears that this is in fact the case.

Laboratory experiments, wildlife studies, and the human DES experience link hormone disruption with a variety of male and female reproductive problems that appear to be on the rise in the general human population—problems ranging from testicular cancer to endometriosis, a condition in which tissue that normally lines the uterus mysteriously migrates to the abdomen, ovaries, bladder, or bowel, resulting in growths that cause pain, severe bleeding, infertility, and other problems.

The most dramatic and troubling sign that hormone disruptors may already have taken a major toll comes from reports that human male sperm counts have plummeted over the past half century, a blink of the eye in the history of the human species. The initial study, done by a Danish team headed by Dr. Niels Skakkebaek and published in the *British Medical Journal* in September 1992, systematically reviewed the international scientific literature on semen analysis performed on normal men since 1938 and based its findings on sixty-one studies involving almost fifteen thousand men from twenty countries in North

America, Europe, South America, Asia, Africa, and Australia. The study excluded men sampled at fertility clinics, who might have particularly low sperm counts, or men who were diseased, and it used only those studies that counted sperm in the same way using light microscopes.

The Danish researchers found that the average male sperm count had dropped forty-five percent, from an average of 113 million per milliliter of semen in 1940 to just 66 million per milliliter in 1990. At the same time, the volume of the semen ejaculated had dropped by twenty-five percent, making the effective sperm decline equivalent to fifty percent. During this period, the number of men with extremely low sperm counts in the range of 20 million per milliliter had tripled, increasing from six percent to eighteen percent, while the percentage with high sperm counts over 100 million per milliliter had decreased.

The study is still meeting with a skeptical response in parts of the medical community. This skepticism recalls similar disbelief at the first news in 1985 that a dramatic hole had developed in the Earth's protective ozone layer over Antarctica. Some scientists then doubted the reports and were even more skeptical that man-made chlorofluorocarbons, known as CFCs, might be responsible. The NASA satellite monitoring network had failed to pick up the ozone loss, which was discovered by British researchers doing measurements from the ground, because those who had programmed the computers receiving the satellite data assumed such large ozone losses were impossible. Similarly, many medical researchers had been incredulous at the first reports of sperm count decline, regarding such a large drop in sperm count across the human population as next to impossible.

Skakkebaek, a specialist in male reproduction and chief of the Department of Growth and Reproduction at the University Hospital in Copenhagen, had been a skeptic himself. Although he had been seeing increasing abnormalities in the male reproductive system, including rising rates of testicular cancer in young men, he had doubted earlier reports of declining human sperm counts over the past two decades. He suspected that they stemmed largely from a bias in the samples, such as the inclusion of men from fertility clinics, and did not, therefore, truly reflect sperm counts among normal men.

His team's own broad review of dozens of sperm count studies from around the world convinced him that a precipitous drop in sperm counts had, indeed, occurred over just two generations—a significant change that is likely, in his view, to have a "negative influence on male fertility." Since such a rapid decline could not be a consequence of genetic changes, the cause must lie in changed living habits or environmental factors.

As these sperm count findings came under scrutiny, critics claimed that flaws in the data made it impossible to draw definitive conclusions. For example, they erroneously criticized Skakkebaek's team for excluding men with abnormally low sperm counts and further objected that the definition of "abnormal" had changed over time. In fact, Skakkebaek and his colleagues made no such exclusion other than to exclude all data from fertility clinics. At the same time, critics offered no data to refute Skakkebaek's conclusion; they simply maintained that he hadn't proven his case beyond doubt.

This debate stimulated other studies, and subsequently at least three independent analyses, undertaken in one instance by another skeptic, have confirmed that sperm counts have dropped. Based on semen samples from 5,440 men, these new studies carried out in France, Belgium, and Scotland have provided additional evidence that the cause is probably environmental.

The new results reveal a striking inverse correlation between the year of birth and the health of men's sperm. The more recently a man was born, the lower the average sperm numbers and the greater the number of sperm abnormalities. For example, the Scottish study conducted by the Medical Research Council's Reproductive Biology Unit in Edinburgh on a sample of 3,729 men found a median sperm count of 128 million per milliliter in semen donors born in 1940 but a median count of only 75 million in those born in 1969.

Belgian researchers, comparing sperm samples from 360 men donated between 1990 and 1993 to earlier samples from 1977 to 1980, found an alarming increase in unhealthy sperm over this sixteen-year period. The percentage of well-formed sperm had declined from 39.6 percent to 27.8 percent, while the percentage of sperm showing normal swimming ability, or motility, dropped from 53.4 percent to 32.8 percent. The authors' conclusion was unusually direct for scientists, who tend toward qualification and understatement: the decline, they warned, "threatens male fertilizing ability."

Most recently, a team of French scientists led by Jacques Auger published a study examining sperm count trends in Paris from 1973 to 1992. Auger had embarked on the analysis because he simply didn't believe the Danish study. To Auger's surprise, his team's analysis lent strong support to the conclusion that male sperm counts have dropped steadily over the twenty-year period.

The French findings are particularly persuasive because the data allowed the researchers to correct for two important confounding variables that might call sperm count results into question: age and abstinence. A man's sperm count generally declines as he gets older, and it drops immediately after sex, recovering within a few days.

The French team was, therefore, able to compare the average sperm counts of thirty-year-old men born in 1945 with thirty-year-olds born seventeen years later in 1962. For those born in 1945 and measured in 1975, sperm counts averaged 102 million per milliliter of semen. The men born in 1962 and measured in 1992 had counts that were only *half* that number—51 million sperm per milliliter on average.

If this downward trend were to continue, the thirty-year-old man in 2005, who was born in 1975, would have a sperm count of roughly 32 million sperm per milliliter—about one-fourth the count of the average male born in 1925.

Declining sperm counts are just one of many signs of trouble. Over the past half century, Skakkebaek noticed, the incidence of testicular cancer and other reproductive abnormalities in men has risen sharply. In Denmark, testicular cancer, a disease of young men, has tripled, and other industrial countries have seen similar trends. "The frightening thing," Skakkebaek notes, "is that the incidence is still increasing in Denmark."

British researchers report a doubling in the number of cases of undescended testicles in England and Wales between 1962 and 1981, and similar increases have been reported in Sweden and Hungary. Men with undescended testicles have a higher risk of testicular cancer and typically have lower sperm counts and more abnormal

sperm. There also appears to be an increasing incidence of the genital defect hypospadia.

In his earlier work, Skakkebaek had slowly accumulated evidence that all these abnormalities stem from developmental errors
that take place in the womb, leading him to suspect they might have
a common cause. He first found signs that prenatal events might
play a role in testicular cancer while exploring the causes of male infertility. Examining tissue samples taken from the testicles of infertile men, he noticed strange-looking cells resembling the fetal germ
cells that evolve during normal development into the testicular cells
that produce and nurture sperm in mature males. The next clue
came when some of the infertile men with aberrant cells later developed testicular cancer. Then he found the same abnormal cells in a
boy with undescended testicles, who also developed cancer a decade
later. It appeared that the abnormal cells had given rise to the testicular cancer and that men suffering from infertility or certain genital
defects were at greater risk of having such cells.

At the same time that Skakkebaek was conducting his review of
sperm numbers, another researcher in Edinburgh, Scotland, was also
puzzling about the increase in male reproductive problems. Richard
Sharpe of the Medical Research Council's Reproductive Biology
Unit was exploring possible explanations for the increase in undescended testicles and the drop in human male sperm counts. He,
too, began to suspect that prenatal events might be responsible.

When Sharpe and Skakkebaek met at a conference and discovered they were thinking along similar lines, they began collaborating.
In May 1993, they published a paper in the distinguished British
medical journal *The Lancet*, proposing that the cause of the falling
sperm counts and increased reproductive abnormalities in men was
increased exposure to estrogens in the womb. They cited the DES
experience and laboratory studies as support for this hypothesis, noting that prenatal exposure to elevated levels of synthetic or natural
estrogens resulted in reduced sperm counts and an increase in the
incidence of undescended testicles, hypospadias, and possibly testicular tumors in male offspring.

Animal studies had also provided an insight into how elevated
estrogen levels in the womb might limit the number of sperm a male

produces in adulthood by inhibiting the multiplication of key cells in the testicles known as Sertoli cells, which orchestrate and regulate the production of sperm. Since each Sertoli cell can support only a fixed number of sperm, the number a male acquires early in life will ultimately limit the number of sperm he can produce as an adult. During male development, the pituitary gland secretes a follicle-stimulating hormone that stimulates the proliferation of the Sertoli cells. Sharpe and Skakkebaek note that studies have shown that elevated estrogen levels suppress the secretion of follicle-stimulating hormone and thereby limit the growth in Sertoli cell numbers.

Of course, one has to bear in mind that many chemicals can undermine male fertility, acting not just before birth but during childhood and adulthood as well. Prenatal exposure to synthetic estrogens is just one of many assaults on male reproductive potential. As we've seen, U.S. Environmental Protection Agency reproductive toxicologist Earl Gray has discovered that some synthetic chemicals act as potent androgen blockers and may pose an even greater threat to males than estrogenic chemicals. Nevertheless, Sharpe and Skakkebaek's theory offers an eminently plausible explanation of how estrogens might be having a major impact on male fertility by disrupting development.

Some reproductive researchers have blamed declining sperm counts on changing lifestyles, noting that smoking and drinking habits and sexual behavior have changed dramatically over the past half century. Studies have shown that heavy drinking and smoking can impair sperm production, and an increased number of sexual partners would expose men to sexually transmitted diseases and genital infections that could also reduce sperm counts.

The latest sperm count studies, however, add weight to the theory that the cause of the precipitous drop in sperm counts is some prenatal event rather than later damage caused by chemical contaminants or bad habits. If sperm counts were dropping because of adult exposure to increased contamination, smoking, or alcohol or because of venereal disease, they would be dropping in older men as well as younger men. The fact that the sperm counts are lower in younger men and correlate inversely with date of birth argues that the damage was done in the womb.

Although the animal studies and the DES experience precisely predict problems such as dropping sperm counts and increasing sperm abnormalities, there is unfortunately no way to prove that prenatal exposure to hormonally active synthetic chemicals has, in fact, caused this alarming deterioration in humans. To do so, Sharpe explains, one would need to know the contaminant levels present in the mothers of the impaired men during their pregnancies decades earlier. Even if, by some chance, it were possible to obtain blood or fat samples taken when these women were pregnant, researchers would still face great uncertainty about which chemicals to measure. Any disruptive effects may be the result of many chemicals working together, as Ana Soto's work has suggested, and the existing list of synthetic estrogen mimics in the environment is almost certainly incomplete. Until all the hormone disruptors are identified, there will be no way to make accurate assessments of an individual's exposure.

Synthetic chemicals have been shown to undermine male fertility even without such obvious signs of damage as dropping sperm counts or sperm abnormalities, particularly if males are exposed at a young age. In rat studies, Dorothea Sager of the University of Wisconsin at Green Bay exposed animals to PCBs early in life through their mothers' breast milk. When the male rats matured, their sperm showed no obvious deficiencies, yet when mated, they were generally unsuccessful in fertilizing eggs, or if fertilized, the eggs failed to develop normally. As with other animals, human infants take in heavy doses of PCBs and other contaminants in breast milk, which exposes them to levels ten to forty times greater than the daily exposure for an adult. Several studies report that infertile men have higher levels of PCBs and other synthetic chemicals in their blood or semen, and one analysis found a correlation between the swimming ability of a man's sperm and the concentrations of specific members of the PCB family found in his semen.

Before biological research gave us an understanding of the complex events involved in reproduction, infertility was invariably "blamed" on the woman who failed to conceive. Today, infertility specialists with the American Fertility Society report that roughly forty percent of the time the failure stems from male problems, notably low sperm counts or poor motility. In the United States, an estimated 5.3 million individuals suffer from infertility, and many resort to an array of invasive and complicated high-tech medical procedures in the effort to have a child. In those cases where the man's sperm is too feeble to penetrate an egg, some specialists are now using an experimental technique that inserts the sperm to allow fertilization. As sperm quality has deteriorated, the medical establishment has responded with costly technological fixes. According to the last official estimate in 1987, the United States was spending $1 billion a year to treat infertility. Those working in the field say that costs have increased substantially since then and may now total as much as $2 billion.

Prenatal exposure to hormone-mimicking substances may also be exacerbating the most common medical problem afflicting aging males: painfully enlarged prostate glands that make urination difficult and often require surgery. Exposure to elevated estrogen levels before birth appears to make males more vulnerable to prostate enlargement later in life. In Western countries, eighty percent of men show signs of this malady by age seventy, and forty to fifty percent of men suffer from gross enlargement of the gland.

In ongoing studies in mice on how hormone levels affect prostate gland development, Frederick vom Saal has found that males exposed prenatally to elevated estrogen levels develop more androgen receptors in their prostates and become permanently sensitized to testosterone. With just a slight increase in estrogen exposure in adulthood, the number of androgen receptors in their prostates will increase an additional 50 percent. This surfeit of receptors induced by excess estrogen makes the prostates in these males permanently hypersensitive to male hormones and vulnerable to prostate enlargement. Treating this benign but debilitating condition with drugs and surgery in the United States costs an estimated $6 billion a year.

Even in males exposed only as adults, chronic low doses of estrogen can have serious consequences for prostate health. In studies with rats, Tufts University researcher Shuk Mei Ho has found that long-term exposure to estrogen can induce prostate cancer.

Over the past two decades, there has been a dramatic increase in this disease, the most common cancer in American men, which accounts for 27 percent of all male cancers. The National Cancer Institute reports a 126 percent rise in prostate cancer from 1973 to 1991—a yearly increase of 3.9 percent. Such incidence figures are adjusted to eliminate the effect of demographic changes such as an increasingly large population of older men. While better detection methods have contributed in part to these skyrocketing prostate cancer rates, specialists in the field believe they are nevertheless seeing a real and alarming increase in this cancer.

Despite more sophisticated methods of early detection, the death rate from prostate cancer continues to rise. From 1980 to 1988 in the United States, reported deaths rose 2.5 percent in white men and 5.7 percent in black men. Health officials attribute the alarming increase in black men to health care inequities, saying that they have less access to screening and to the more advanced and costly treatments for the disease.

The bare facts are compelling. The animal studies show links between elevated estrogen levels and prostate disease. As human exposure to synthetic estrogens has increased over the past half century, so has the incidence of prostate disease. Exploring the role of hormonally active synthetic chemicals in these hormonally driven diseases should be a top research priority.

The DES experience and animal studies also suggest links between hormone-disrupting chemicals and a number of reproductive problems in women, notably miscarriages, tubal pregnancies, and endometriosis.

In tubal, or ectopic, pregnancies, the fertilized egg begins developing in the fallopian tubes rather than the uterus, a dangerous event that can be life-threatening or cause the loss of the fallopian tube. Repeated tubal pregnancies can result in infertility. While some medical specialists have blamed this increase on higher rates of sexually transmitted diseases that can permanently damage the fallopian tubes and other parts of the reproductive tract, the DES history suggests that hormone-disrupting chemicals may be a significant factor as well. DES daughters suffer three to five times more tubal pregnancies than unexposed women, and there is evidence that the rate is soaring in the general population, too. A 1990 study in Wisconsin found that the rate of ectopic pregnancies increased 400 percent between 1970 and 1987.

Endometriosis, a poorly understood disease that affects an estimated 5.5 million women in the United States and Canada and is a prominent cause of infertility, also appears to be increasing and showing up more than ever before in very young women.

It is difficult, however, to document long-term trends or determine the precise number of women afflicted with the disease. Many cases go undiagnosed, and accurate diagnosis requires an invasive procedure called a laparoscopy. Specialists in the field believe that the prevalence of the disease has increased greatly since World War II. The National Institute of Child Health and Human Development estimates that endometriosis afflicts ten to twenty percent of women of childbearing age in the United States. Prior to 1921, there were only twenty reports of the disease in the worldwide medical literature.

After years of debate about the cause of the disease, which appears to be associated with some alteration of the immune system, a recent study with rhesus monkeys has provided the first solid clue. A study on a colony of rhesus monkeys found that the animals spontaneously developed endometriosis a decade after their first exposure to the pollutant dioxin. The greater a female monkey's exposure to dioxin, the greater the severity of the disease. German researchers reported in a recent study that women with endometriosis have higher levels of PCBs in their blood than women who do not suffer from the disease. Dioxin and dioxinlike PCBs are known to affect the immune system as well as many parts of the endocrine system. Sparked by these findings, additional studies are under way, including an effort by the National Institute of Environmental Health Sciences to assess the blood concentrations of dioxin, furans, and PCBs in women with endometriosis.

Animal studies also clearly signal that exposure to certain synthetic chemicals, such as PCBs, increase the risk of miscarriage, a link that has been reported in human studies as well. Concentrations of the hormone progesterone, which is necessary to maintain pregnancy, must remain high throughout pregnancy to avoid the loss of the developing embryo. Researchers studying women who suffer miscarriages report that they have higher than average PCB levels in their bodies compared to women who have normal pregnancies. The studies in rats and mice indicate that PCBs cause a reduction in progesterone by accelerating its breakdown by the liver.

But by far the most alarming health trend for women is the rising rate of breast cancer, the most common female cancer. Despite the flurry of recent publicity about a "breast cancer gene," researchers estimate that only about five percent of breast cancers are a result of inherited genetic susceptibility. A large proportion stem, therefore, from other, noninherited factors. As a general principle, the risk of breast cancer is linked to a woman's total lifelong exposure to estrogen. Early menstruation and late menopause, for example, will increase a woman's risk.

Because total estrogen exposure is the single most important risk factor for breast cancer, estrogenic chemicals, which would add to this lifelong exposure, are an obvious suspect when searching for the cause of rising rates over the past half century. Since 1940, when the chemical age was dawning, breast cancer deaths have risen steadily by one percent per year, in the United States, and similar increases have been reported in other industrial countries. Such incidence rates are adjusted for age, so they reflect genuine trends rather than demographic changes such as a growing elderly population. Among American women forty to fifty-five years of age, breast cancer is now the leading cause of death.

From 1980 to 1987, the number of breast cancer cases reported in the United States jumped by thirty-two percent, raising the specter of a breast cancer epidemic, although one study concluded that much of this increase may be due to the increasing use of mammograms for breast cancer detection. Whether this dramatic rise is real or in some part an artifact of improved detection, the steady rise in breast cancer deaths over the past two generations is by itself a cause for concern. Fifty years ago, a woman ran a one in twenty risk of getting breast cancer. One in eight women in the United States today will get breast cancer in her lifetime.

The most notable increase in breast cancer has been in postmenopausal women with estrogen-responsive tumors—tumors that are rich in estrogen receptors and proliferate when exposed to estro-

gen. In patients over fifty, researchers report both an increasing proportion of cases of estrogen-responsive breast cancer and an increasing density of estrogen receptors within these tumors. This finding is
based on an analysis of more than eleven thousand breast tumor
specimens sent from hospitals across the United States to the University of Texas Health Science Center. In reporting their results in
the journal *Cancer,* the team suggested that this increase may reflect
a change in the hormonal events that promote breast cancer development, such as the onset of menstruation or pregnancy, or exposure to estrogens other than those produced within the body.

In 1993, a group of researchers working on the puzzle of rising
breast cancer rates proposed the theory that hormonally active synthetic chemicals are causing the rise in breast cancer incidence and
deaths among older women by increasing their overall estrogen exposure. The group, which includes a variety of researchers from various institutions in the United States hypothesizes that synthetic
chemicals may do this directly by acting as estrogen mimics or indirectly by altering the way the body produces or metabolizes estrogen.
They also theorize that prenatal exposure to estrogens may predis-
pose a woman to breast cancer later in life through an "imprinting"
process that sensitizes her to estrogen exposure. Such an imprinting
would be similar to the prenatal sensitization process that vom Saal
has discovered in the male prostate gland.

Studies by H. Leon Bradlow and his colleagues at the Strang
Cornell Cancer Research Laboratory have found evidence for one
way that synthetic chemicals may increase breast cancer risk—by altering the way the body processes its own estrogen.

Bradlow, who has been investigating links between estrogen
metabolism and cancer, has discovered that the body can process the
form of estrogen known as estradiol in two chemically different
ways—one that might be called a "good" estrogen pathway, which
produces a weak form of estrogen, and the other "bad," because
it produces a potent estrogen that can increase the risk of cancer. In
experimental studies, he has found that various factors can influence
which of these two mutually exclusive pathways the body uses. For
example, studies by Bradlow and his colleagues have found that a
substance found in broccoli, cauliflower, brussels sprouts, and other

members of the cabbage family, indole-3-carbinol, reduces cancer risk by pushing estrogen metabolism toward the good pathway.

In recent studies, however, the Cornell researchers report that hormonally active chemicals have the opposite effect—they push estrogen metabolism toward the bad pathway and increase cancer risk. The experiments, which exposed breast cancer cells in test tubes to synthetic chemicals including DDT, PCBs, endosulfan, kepone, and atrazine, found that all these chemicals have a "profound effect" and greatly increase the production of a bad form of estrogen.

"Our data show that a wide variety of pesticides and related compounds clearly have effects on estrogen metabolism that would act in the direction of increasing breast cancer and endometrial cancer risks," Bradlow says.

Researchers have been exploring possible links in other ways as well, including analyzing body concentrations of synthetic chemicals in women who develop breast cancer. A Canadian study involving a small number of women found significantly higher levels of DDE, a DDT breakdown product, in women with estrogen-responsive tumors, supporting the hypothesis that exposure to hormonally active chemicals may affect the incidence of hormone-responsive breast cancer.

Two other studies collected and stored blood samples from large numbers of healthy women without any signs of cancer. When some of the women later developed cancer, researchers analyzed the blood looking for differences in exposure to DDT and PCBs between those who developed cancer and those who did not. The New York University Women's Health Study, whose study group was composed primarily of Caucasian women, found that those with the highest levels of DDE had a fourfold greater risk than those with the lowest levels. But the second study done in California with a larger group of women, which included Caucasian, African-American, and Asian women, found no link between breast cancer and DDE levels overall.

Whatever the design, the studies to date have several shortcomings that make such inconsistent results not at all surprising. The contribution of synthetic chemicals to estrogen exposure may come from many different chemicals, some of them exceedingly persistent, such as PCBs, and others that are not persistent and leave no telltale evidence of exposure in blood or body fat. The studies to

date have looked only at a handful of well-known persistent compounds such as DDE and PCBs.

Also, studies have usually treated all 209 chemicals in the PCB
family as one, even though various members of this chemical family
have completely different and in some cases opposite biological effects. Some are estrogen mimics, while the dioxinlike compounds can
act as estrogen blockers. Moreover, the mix of PCBs found in humans
varies from individual to individual, depending on their diet and other
exposure. Teasing out any correlation between PCBs and breast cancer is going to require treating the PCBs as individual chemicals.

The discovery that estrogenic chemicals lurk in plastics, canned
goods, and detergent breakdown products also suggests that significant exposure may result from chemicals other than the usual suspects. Studies have not screened human tissue for less well-known
estrogen mimics such as bisphenol A or nonylphenol, even though
they could conceivably be contributing in a major way to estrogen
exposure from synthetic chemicals. Some researchers, including Ana
Soto and Carlos Sonnenschein, are now exploring ways to separate
out the natural estrogen found in women's blood from foreign estrogens in an effort to determine how much overall estrogen exposure
comes from synthetic chemicals.

Because of our poor understanding of what causes breast cancer
and significant uncertainties about exposure, it may take some time to
satisfactorily test the hypothesis and discover whether synthetic chemicals are contributing to rising breast cancer rates. In part because
of political pressure from breast cancer advocacy groups on the U.S.
Congress, federal funding agencies are now directing more money
toward exploring this important question. The National Cancer Institute is currently funding a 6-million-dollar four-year study to investigate the connection between breast cancer and environmental
exposure to several hormonally active synthetic chemicals. This effort,
known as the Northeast/Mid-Atlantic Study, is focusing particularly
on the possible connection between synthetic pollutants and the elevated breast cancer risk faced by women living in this region.

Among cancer victims in the United States, breast and prostate
cancer are two of the four biggest killers. Hormones play a major role
in both these cancers, and the rates of incidence of both continue to

climb upward not only in the United States but in most other countries as well. The highest rates of breast cancer deaths are in Western Europe, followed by the United States, but the fastest increase in deaths from the disease is now occurring in Eastern Europe and East Asia. For prostate cancer, the highest death rates occur in the Nordic countries, while East Asia, which has a very low incidence, is seeing the greatest increase. The quest to discover the role of hormone-disrupting chemicals in these cancers deserves higher priority than the quest for hereditary breast and prostate cancer genes because research aimed at environmental factors offers the hope of finding ways to prevent these devastating diseases in the vast majority of victims.

Our fears about toxic chemicals have typically centered around cancer and other physical illnesses. But as one surveys the scientific literature, it becomes quickly apparent that physical disease or visible birth defects may not be the most immediate danger. Long before concentrations of synthetic chemicals reach sufficient levels to cause obvious physical illness or abnormalities, they can impair learning ability and cause dramatic, permanent changes in behavior, such as hyperactivity. Save for a few compounds such as PCBs, we know virtually nothing about the hazards posed to thinking and behavior by the thousands of synthetic chemicals in commerce.

What little we do know about those few chemicals that have been studied has alarming implications. Both animal experiments and human studies report behavioral disorders and learning disabilities similar to those reported with increasing frequency among school children across the nation. In the United States, an estimated five to ten percent of school-age children suffer from a suite of symptoms related to hyperactivity and attention deficit that make it difficult for them to pay attention and learn. Countless others experience learning problems ranging from difficulties with memory to impaired fine motor skills that make it harder to hold a pen and learn how to write.

Scientists still do not have a complete understanding of how PCBs impair neurological development in the womb and early in life, but emerging evidence suggests that the ability of PCBs to cause brain damage stems in part from disruption to another component of the endocrine system, thyroid hormones.

Extensive research on the developing brain and nervous system has found that thyroid hormones help orchestrate the elaborate step-by-step process that is required for normal brain development. As touched on in Chapter 3, these hormones stimulate the proliferation of nerve cells and later guide the orderly migration of nerve cells to appropriate areas of the brain. The brain and nervous system, like other parts of the body, pass through critical periods during their development both in the womb and in the first two years of life. When thyroid levels are too high or too low, this development process will go awry and permanent damage will result, which can range from mental retardation to more subtle behavioral disorders and learning disabilities. The precise nature of the damage done by abnormal thyroid levels will depend on the timing and the extent of the disruption.

It has long been recognized that acute thyroid deficiency during pregnancy can cause profound mental retardation, but thyroid researcher Susan Porterfield, an endocrinologist at the Medical College of Georgia, notes that few have considered the more subtle effects of less severe thyroid disruption during the development of the brain and nervous system—disruption that can occur naturally or be the result of hormone-disrupting chemicals in the environment.

PCBs and dioxin affect the thyroid system in diverse, complex, and as yet incompletely understood ways. Some analyses indicate they may mimic or block normal hormone action perhaps by binding to the thyroid receptor. Other data suggest they may even increase the number of receptors present to receive the hormone signals. They also seem to act particularly on T4, the form of thyroid hormone that is critical to prenatal brain development. Researchers Daniel Ness and Susan Schantz of the University of Illinois have established that two PCBs commonly found in human tissue and breast milk—PCB-118 and PCB-153—reduce T4 levels in rats exposed prenatally. These compounds also compete more powerfully than natural hormones for binding to a carrier protein called transthyretin, which transports T4 to brain cells.

In a June 1994 article in the journal *Environmental Health Perspectives*, Porterfield outlined her theory that "very low levels" of PCBs and dioxins—levels well below those generally recognized as toxic—can alter thyroid function in the mother and the unborn baby and thereby impair neurological development. Like Sharpe and Skak-kebaek, Porterfield cites evidence showing that skewed hormone levels in the womb can cause permanent damage—in this case, learning disabilities, attention problems, and hyperactivity.

The emerging evidence linking PCBs to thyroid disruption and neurological damage is especially worrisome because PCBs are a persistent, ubiquitous contaminant whose levels had dropped initially in human tissue but have held steady in recent years, even though most of the industrial countries stopped making them more than a decade ago. In the former Soviet Union, production of PCBs did not stop until 1990.

Based on the concentrations in breast milk fat of PCBs, some have estimated that at least five percent of babies in the United States are exposed to sufficient levels of contaminants to cause neurological impairment. But while the most extensive data on neurological effects concern PCBs, it is important to stress that PCBs are not by any means the only culprit. Many other synthetic chemicals act on thyroid hormones as well, adding to the concern. The thyroid system is one of the most frequent targets for synthetic chemicals, according to Linda Birnbaum, who heads the environmental toxicology division at the U.S. Environmental Protection Agency's Health Effects Research Laboratory. With the possibility for multiple assaults on the thyroid system, the hazards to the developing brain may be considerable.

Laboratory animals exposed to PCBs in the womb and early in life commonly show behavioral abnormalities as adults. The offspring of some mice fed relatively low doses of PCBs develop a "spinning syndrome" in which they constantly circle their cage. Other exposed mice, while showing no overt behavioral abnormalities, have depressed reflexes and learning deficits. Rats exposed in the womb make more errors in running a maze and have greater difficulty learning how to swim, perhaps reflecting motor impairment. In rhesus monkeys, researchers also find that exposure to PCBs in the womb and through breast milk causes motor impairment as well as deficits in memory and learning. The greater the PCB exposure, the greater the number of errors the monkeys made in learning tasks designed to test their cognitive skills.

But the most obvious and thus most reported behavioral sign of neurological damage found in laboratory animals exposed to PCBs in the womb and early life is hyperactivity, which has shown up in rats, mice, and monkeys. Although one might expect that extrapolation from animal studies to humans would be less reliable when considering questions of behavior and cognition, some striking parallels between animal and human effects appear in these neurological studies as well. In her paper, Porterfield notes that this problem occurs at higher frequency in the children born to women who had abnormally low levels of thyroid hormones during pregnancy.

Much of our knowledge about the impact on humans comes from studying those individuals exposed in accidents. One major long-term study involves the children born to women in Taiwan who in 1979 consumed cooking oil accidentally contaminated with high levels of PCBs and furans. Some of the 128 children studied were in the womb during the time their mothers actually consumed the contaminated oil. Others were conceived and born after the period of contamination had ended, so their exposure came from residual contamination within their mothers. In a series of examinations and tests conducted on these children between 1985 and 1992, a team headed by Yue-Liang L. Guo of Taiwan's Department of Occupational and Environmental Health found that the children suffer from an array of problems—physical and neurological—that had been predicted from animal studies.

As some of these children approached puberty, researchers noted abnormal sexual development in the males, paralleling one of the most striking effects recorded in wildlife literature. Like the alligators in Lake Apopka, these boys have significantly shorter penises than unexposed boys of the same age.

These Taiwanese children also show permanent impairment in their motor and mental abilities and behavior problems including higher than normal levels of activity. Tests have repeatedly found signs of retarded development, and the children scored five points lower on intelligence tests than similar unexposed children. Guo and his colleagues believe that these children score lower in IQ tests because they suffer from attention deficits and are unable to think as quickly as their unexposed peers.

Fortunately, few people are exposed to the intense contamination suffered by victims of this unfortunate accident in Taiwan. Two studies done in the United States have attempted to discover whether children suffer neurological damage when exposed through their mothers to the normal range of contamination encountered in the environment. Both reported signs of impaired neurological development, which might not be evident to the parents but could be detected through specialized tests.

The first study, done in the early 1980s by psychologists Sandra and Joseph Jacobson of Wayne State University, enlisted new mothers in Michigan who had eaten Great Lakes fish, which contain significant levels of PCBs and numerous other chemical contaminants. Despite the contamination levels, state fish and game agencies in the Great Lakes region continue to stock salmon, lake trout, and other game fish, and sport fishing remains a 3- to 4-billion-dollar industry. Signs posted by state health officials at some fishing areas warn that eating salmon and lake trout may be hazardous to health, but many fishermen and their families continue to eat what they catch. The women in the fish-eating group in this study had all eaten two or three fish meals a month in the six years before becoming pregnant, although some had eaten no fish during pregnancy. Since PCBs are persistent, these women accumulated them in their body fat and then passed the PCBs on to their babies through the placenta and through breast milk.

Differences between the children of fish-eaters and nonfish-eaters were evident immediately at birth. The higher the mother's consumption of Lake Michigan fish, the lower the birth weight and the smaller the head circumference of her baby. A series of tests done at birth and at intervals afterward also found persistent evidence of neurological impairment. The Jacobsons cannot be certain, however, that PCBs are solely responsible for the effects seen in the children born to these women because their mothers were exposed to so many other chemicals as well.

Among the more than three hundred children tested in this study, those whose mothers had eaten greater quantities of fish showed subtle signs of damage, including weak reflexes and more jerky, unbalanced movements as newborns. In later testing at seven months of age, the Jacobsons found signs of impaired cognitive function based on a test in which a child is shown two identical pictures of human faces posted on a board. After a period of time, the researcher will remove the board, replace one of the pictures with a new face, and show the display to the child again. A child normally recognizes the new face and spends more time looking at it rather than the face the child had seen before. The higher the PCB levels in the mother, the less time an infant spent looking at the new picture. Children with lower scores on this test tend to perform more poorly on intelligence tests during childhood. When the children were tested again at four years of age, the children of women with the highest PCB levels had lower scores in verbal and memory tests.

The second study done in North Carolina involved neurological tests on 866 infants and compared their performance to the levels of PCBs detected in their mother's breast milk—an indication of their prenatal exposure as well as their exposure after birth. The more highly exposed infants showed weaker reflexes, and in follow-up studies done at six and twelve months of age, they still showed poor performance on tests for gross and fine motor coordination. This study did not include tests to assess thinking and memory skills.

At the State University of New York at Oswego, which sits on the shore overlooking Lake Ontario, a team made up of psychologists and a physician is now extending the groundbreaking research done by the Jacobsons into differences between children of those who do and do not eat Lake Ontario fish. The effort includes a human study on the children of fish-eating women and parallel laboratory research with rats who are being fed Lake Ontario fish. If the human and rat studies find the same changes in behavior, it will indicate that the results of rat studies can be generalized to humans. Since the rat studies done by psychologist Helen Daly are showing evidence of behavioral changes in adult rats, the research team, which also includes Edward Lonky, Thomas Darvill, Jacqueline Reihman, and Joseph Mather, Sr., is testing the parents in the human study as well as the children. A similar epidemiological effort is also underway in the Netherlands, where researchers are exploring the link between PCB exposure, thyroid levels at birth, and later behavioral and cognitive problems in the children of mothers with high PCB levels.

Daly's studies with rats fed Lake Ontario salmon have added

another dimension to the growing literature on the impact of synthetic chemicals on thinking and behavior and new questions about possible effects on humans. Her early work had focused on learning in normal rats, particularly the role of frustration in the learning process. But in the early '80s, she and her colleagues began to wonder about the possible effect of eating contaminated fish. The question was literally inescapable in Oswego, for as the state began to stock game fish, the city at the mouth of the Oswego River on Lake Ontario became a booming sports-fishing center. In autumn, fishermen, who often come from great distances, line up shoulder to shoulder along the riverbank angling for one of the huge salmon heading up the river to spawn. The town has even built special cleaning stations where fishermen can have their fish dressed for a small fee. In conversations with fishermen and their wives, Daly and her colleagues learned that many families were filling their freezers with fish and eating significant amounts through the winter.

A number of researchers at the university begin to feed laboratory rats with Lake Ontario fish to see if it affected them in any way. One of Daly's colleagues, David Hertzler, found signs of changed behavior in rats after twenty days on a diet that was thirty percent salmon. When put through a standard test, the rats showed decreased activity. The finding was intriguing, but since many factors can cause decreased activity, it did not give much insight into how the contamination was affecting the animals. Daly then embarked on a series of learning experiments to try to discover why they were becoming less active.

"We thought that toxic chemicals should make them dummies," she recalls. "That's what you would expect."

What she found was totally contrary to her expectations. There was no evidence of severe learning deficits, but there was a dramatic change in their behavior. In the years since then, Daly has conducted a series of different experiments trying to pinpoint and characterize the nature of this change. The studies all compare the behavior of rats fed on Lake Ontario salmon to rats fed with Pacific Ocean salmon that contain much lower levels of contamination or to a second control group that is not fed salmon. Over and over again, she has seen a consistent pattern in the results.

The two groups of rats show no difference in behavior as long as life is pleasant and uneventful, but as soon as they are confronted by any sort of negative event, great differences are apparent. In every instance, the rats who have eaten Lake Ontario salmon show a much greater reaction than those who have dined on Pacific salmon or rat chow. Daly describes them as "hyper-reactive" even to mildly negative situations.

If the contaminants have similar effects on humans as on rats, Daly says, "every little stress will be magnified." She thinks that some of the children in the Jacobsons' study showed reactions similar to those she has seen in the rats. When the children were tested at four years of age, seventeen of them refused to cooperate during at least one test, all of them children of mothers with the highest levels of PCBs in their milk. She thinks this is probably an overreaction to the mildly frustrating experience of taking such tests.

Daly's findings are unusual in other respects, for she saw this behavioral change, not only in those exposed during critical development stages in the womb but also in adult rats fed Lake Ontario salmon. She also found such changes in the offspring of these adults and in the second generation as well. In studying these transgenerational effects, Daly fed the fish only to the grandmothers before and during their pregnancies and while they were feeding their pups. The offspring got the contaminants only through the mother's placenta and through breast milk. Daly fed none of the female offspring fish at any point in their lives, yet she still saw behavioral changes in the pups they produced. Her studies suggest that contaminants taken in by a mother can somehow have effects that reach across two generations and affect grandchildren as well as immediate offspring. Other researchers are beginning to repeat Daly's experiments, carefully following the same procedures to see whether they get the same results.

In May 1995, the Oswego team announced the initial results from their ongoing human study, reporting behavioral and neurological differences in the children of women who had eaten Lake Ontario fish. In this new study, babies whose mothers had eaten surprisingly modest amounts of Lake Ontario fish—the equivalent of forty pounds or more of salmon over a lifetime, not just during pregnancy—showed a larger number of abnormal reflexes, expressing greater immaturity in a lower autonomic response score, and poor habituation to repeated disturbances.

The habituation assessment, which was not a part of the Jacobson study, looked at a sleeping baby's response to awakening by a bell, a rattle, or a light shining in its eyes. Such stimulation initially startles the baby, but if he or she is awakened repeatedly, the startle response should diminish and eventually disappear. The babies of women who had eaten no Lake Ontario fish grew accustomed, or habituated, to the disturbance most quickly. In contrast, children of heavy fish-eaters habituated poorly, reacting much more negatively to repeated disturbances.

"This follows beautifully," Daly said, speaking of the similarities emerging between the human studies and her earlier work with rats. Like the human infants, she notes, "my rats [fed on Lake Ontario salmon] are also more reactive to unpleasant events." The findings from this first large-scale replication of the Jacobsons' study are also consistent with behavioral differences noted in that earlier effort, although this study did not find the physical differences reported by the Jacobsons such as smaller head circumference and lighter birth weight in babies of heavy fish-eaters. Daly and her colleagues believe this is evidence that contaminants will affect behavior before they reach levels high enough to have a measurable physical impact on a baby. Daly is encouraged by evidence that rodent studies can provide an early warning about possible behavioral effects in humans.

Although the Jacobson study found correlations between neurological symptoms and PCBs, Daly doubts that they are the only chemical involved in the changes the Oswego team is seeing. Estimates for the number of toxic chemicals released in the Great Lakes basin range as high as twenty-eight hundred. How many of them find their way into the salmon and the people who eat them is anybody's guess. In such a toxic stew, she says, it would not be surprising if a number of chemicals are acting in an additive or synergistic fashion.

Reports linking wildlife contamination with such unexpected behaviors as females sharing nests and feminized males inevitably raise questions about human parenting and sexual choice. Could hormone disruption alter these human attributes? The science on this is slender indeed. While emerging evidence suggests that variations in sexual preference may stem from differences in biology, scientists have only a dim understanding of the factors involved.

In 1991, Dr. Simon LeVay published data in *Science* on his discovery of differences in the brain structure of homosexual and heterosexual men. LeVay's work, including his recent book, *The Sexual Brain*, supports the theory that sexual behavior is biologically based and strongly influenced by exposure to hormones during the period when the fetal brain undergoes sexual differentiation.

Other scientists' work supports this interpretation. For example, some studies in DES daughters find that they have higher rates of homosexuality and bisexuality than do their sisters who were not exposed to this synthetic estrogen before birth. Unfortunately, no credible studies exist on the effects of this synthetic estrogen on DES sons.

This body of work would indicate, in principle, that chemicals interfering with hormonal messages at crucial times in fetal development could alter sexual choice. Current science tells us little more. Studies have shown that disrupting hormone levels can sometimes masculinize and sometimes feminize. Thus if endocrine disruption does influence sexual choice, it could conceivably cut both ways, causing a person who might by nature have been homosexual to develop as heterosexual, or causing some who were destined to be heterosexual to become homosexual. One must also remember that diversity in sexual choice has been part of the human experience for millennia, since long before chemical contaminants became widespread. There is no indication that patterns of human sexual choice have changed since synthetic endocrine-disrupting chemicals entered into commerce. While gays and lesbians have become more visible thanks to social trends that have enabled them to participate as full members of society, the best studies available indicate the proportion of homosexuals and heterosexuals in the population has remained constant.

Human sexual orientation is no doubt a complex phenomenon, just as is most human behavior. We doubt any single factor— nature, endocrine disruption, or nurture—will prove to be its sole determinant.

* * *

At the moment, there are more questions than answers about the impact of hormone-disrupting chemicals on humans.

Even if damage is apparent and documented, however, it will never be possible to establish a definitive cause-and-effect connection with contaminants in the environment. Although we know that every mother in the past half century has carried a load of synthetic chemicals and exposed her children in the womb, we do not know what combination of chemicals any individual child was exposed to, or at what levels, or whether he or she was hit during critical periods in their development when relatively low levels might have significant lifelong effects. This is a common and inescapable dilemma in trying to assess the delayed effects of environmental contamination. We also face the problem of having no genuine control group of unexposed individuals for comparative scientific studies. The contamination is ubiquitous. Everyone is exposed to some degree. One of the sad ironies is that researchers discovered the high levels of contamination among people in remote Inuit villages while looking for a less exposed control group.

For these reasons, those who demand such definitive "proof" before reaching a judgment are certain to be waiting an eternity. In the real world, where humans and animals are exposed to contamination by dozens of chemicals that may be working jointly or sometimes in opposition to each other and where timing may be as important as dose, neat cause-and-effect links will remain elusive.

The tobacco industry disingenuously used arguments about the lack of a proven cause-and-effect link between smoking and lung cancer in humans, knowing full well it is impossible to obtain such evidence in humans without subjecting them to controlled laboratory experiments. But after delays, health officials have moved ahead with warnings on cigarettes, limits on cigarette advertising, and efforts to stop exposure to cigarette smoke in public places. They have done so based on evidence of a correlation between smoking and lung cancer in humans backed up by controlled laboratory experiments with animals that do have the ability to demonstrate such a cause-and-effect relationship.

It will take a similar approach to tackle the problem of hormone disruptors, but it will be vastly more difficult than untangling the web of cause and effect for tobacco. Given the nature of the contamination, it is important to recognize at the outset that those responsible for safeguarding human health will have to act on information that is less than perfect.

As we wrestle with the question of how much chemical contaminants are contributing to the trends and societal patterns we see—in breast cancer, prostate disease, infertility, and learning disabilities—it is important to keep one thing in mind. Scientists keep finding significant, often permanent effects at surprisingly low doses. The danger we face is not simply death and disease. By disrupting hormones and development, these synthetic chemicals may be changing who we become. They may be altering our destinies.

BEYOND CANCER

EARLY IN HER DETECTIVE WORK, THEO COLBORN STUMBLED ONTO A long-forgotten study published in the Proceedings of the Society of Experimental Biology and Medicine in 1950—the first warning in the scientific literature that synthetic chemicals could have the inadvertent effect of disrupting hormones. The paper by two Syracuse University zoologists, Verlus Frank Lindeman and his graduate student Howard Burlington, described how doses of DDT prevented young roosters from developing into normal males and even suggested that the pesticide was acting as a hormone. So the first bizarre, frightening evidence of hormone havoc had surfaced soon after the chemical age swept into American life at the end of World War II like a tsunami. Colborn posted the Burlington and Lindeman paper above her desk—a reminder of the slow acceptance of new ideas.

How did we miss so many warning signs, and for so long?

Colborn's own experience with the Great Lakes provides part of the answer. Our obsession with cancer blinds us to other dangers. There is a strong tendency, seen again and again in this story, to overlook or ignore important new evidence that does not fit into reigning concepts about how things work and what is important—a strong tendency to turn a deaf ear.

If Burlington and Lindeman's study sank into oblivion, it certainly wasn't because their findings were subtle or hard to understand. The Syracuse team decided to explore the effects of long-term poisoning from DDT by injecting the pesticide into forty young roosters for a period of two to three months. As they watched what happened to the white leghorns, they must have puzzled about this most peculiar poison. The daily doses of DDT didn't kill the roosters or even make them sick. But it certainly did make them weird. The treated birds didn't look like roosters at all; they looked like hens.

As young cockerels mature, tall red combs blossom on their heads and luxurious cherry-colored skin folds called wattles burgeon at their necks—hallmarks of rooster masculinity. The birds dosed with DDT, however, failed to develop in the expected way. Even as adults, their combs and wattles remained pale and stunted, measuring just one-third the size of those displayed by untreated roosters raised in the zoology lab for the purpose of comparison. When the researchers examined the birds' testicles, the findings were even more startling. The sex organs had grown to only eighteen percent of normal size. From all appearances, the birds had suffered chemical castration.

Decades would pass before scientists would begin to understand exactly how DDT had altered the sexual destiny of the young roosters. But in discussing the "interesting effects" of their experiment, the Syracuse team teased out the ominous implications with a remarkable prescience. They suggested—in 1950—that DDT may be exerting an "estrogenlike action," that it was acting like a hormone.

The study provided alarming evidence of the power of a synthetic chemical to derail sexual development, but this warning went unheeded. Such findings simply did not fit into prevailing views about pesticide hazards, which had been shaped by the previous generation of poisons—many of them acutely toxic arsenic-based compounds that left dangerous residues on fruits and vegetables that had sometimes killed people outright. Based on this prewar experience, public health officials in 1950 thought in terms of classical poisoning and judged chemicals safe if they did not cause death or obvious disease in those exposed to high concentrations, such as farmers.

By this measure, DDT, which came onto the civilian American market in 1946, was a remarkably safe product. Within a year, this "miracle" pesticide was in widespread agricultural use in the United States. Between 1947 and 1949, chemical companies had poured $3.8 billion into production facilities aimed at a vast new market for synthetic pesticides. Sales of DDT alone, which totaled $10 million in 1944, soared to more than $110 million by 1951. On farms, in homes and gardens, and along suburban streets as a part of mosquito control efforts, DDT was spread and sprayed with a casual abandon that is now hard to imagine.

By the 1960s, new fears about the health effects of pesticides were coming to the fore—fears about cancer that would dominate the public debate about toxic chemicals, scientific research, and government regulation for the next three decades. Reflecting this shift in public consciousness, Rachel Carson focused on cancer in her chapter on pesticides and human health in *Silent Spring*, even though she had found ample evidence of other hazards in her exploration of the scientific literature. The "Fable for Tomorrow" that opens the book paints a chilling picture of reproductive failures. "On the farms the hens brooded, but no chicks hatched. The farmers complained that they were unable to raise any pigs—the litters were small and the young survived only a few days."

Carson evidently read the Burlington and Lindeman study because she refers to their results, but without citation or mention of their names. Although it is now clear that their hypothesis about DDT's hormone actions sheds light on some of the wildlife symptoms detailed in the opening chapters of *Silent Spring*, Carson never pursued the clues that pesticides might impair reproduction by disrupting hormones. This is particularly intriguing, since she clearly recognized that "something more sinister" than straightforward poisoning was occurring—"the actual destruction of the bird's ability to reproduce." Somehow this thread was lost, perhaps because of the state of scientific understanding about the endocrine system at the time, perhaps because Carson, who was herself suffering from breast cancer, was as preoccupied with the dreaded disease as her readers. In the later chapters of *Silent Spring*, the reproductive theme so prominent early in the book disappears as Carson turns her full attention to concerns that synthetic pesticides cause genetic mutations and cancer.

It is only in this context that Carson touches on the idea that pesticides might somehow interfere with hormone levels and thereby abet cancers of the reproductive system. This might occur indirectly, she argues, through damage to the liver, which is easily harmed by synthetic chlorinated compounds. The liver plays a key role in maintaining hormone balance by breaking down estrogen and other steroid hormones to allow for their excretion. If impaired liver function slowed this breakdown process, she speculated, it could lead to "abnormally high estrogen levels." Carson was correct, medicine now acknowledges, in linking overall estrogen exposure to these cancers and in recognizing that synthetic chemicals can disrupt hormones by impeding normal liver processes. Now, thirty years later, the inquiry into the role of DDT and other synthetic chemicals in breast cancer has been renewed following the discovery in breast tumors of DDT, of its breakdown product DDE, of PCBs, and of other synthetic chemicals and the recognition that some synthetic chemicals are indeed hormonally active, as Burlington and Lindeman first suggested more than four decades ago.

Cancer holds a special dread in our culture. If Rachel Carson helped publicize suspected links between rising cancer rates and increasing use of synthetic pesticides, other forces added momentum to the emerging cancer paradigm. In June 1969, the National Cancer Institute completed animal tests prompted by *Silent Spring* and did find a higher incidence of liver tumors in mice exposed over a long period to low levels of DDT. Within a week, President Richard Nixon received a petition from seventeen congressmen who asked for a ban on DDT on the grounds that it caused cancer. In 1971, President Nixon declared an all-out War on Cancer. Given the temper of the time and new evidence that pesticides like DDT might cause cancer, environmentalists seeking to ban DDT began to frame the battle around human health risks rather than the wildlife and environmental concerns put forward in earlier phrases of the long campaign. In his 1972 decision to restrict most uses of DDT, U.S. Environmental Protection Agency Administrator William Ruckelshaus gave equal prominence to possible human cancer risks and to the adverse impacts on fish and wildlife.

Just as cancer holds a special place in our fears, it also commands a special place in federal regulation. It has defined and driven the EPA's regulatory process for toxic chemicals for more than two decades, mainly because the agency uses different assumptions for assessing cancer risk than it does when considering other risks. For noncancer hazards such as reproductive and developmental damage, the agency assumes that a chemical may pose no hazard in low concentrations beneath some threshold level. But when it is a question of cancer, the EPA turns to a linear model, which assumes that no level is safe. Even the tiniest dose of a chemical is presumed capable of causing cancer.

The federal appeals court had reinforced such a regulatory bias by a series of rulings on pesticides in the early 1970s, beginning with a 1970 case concerning federal tolerance levels for DDT residues on food. The Environmental Defense Fund, which brought the action, based its case on the Delaney clause, a 1958 amendment to the Federal Food, Drug, and Cosmetic Act, which prohibits the use of any food additive that has been shown to cause cancer in laboratory animals. In its opinion, the court decided that this law applies to pesticide residues as well and that the carcinogenicity of these residues has to be considered in setting tolerance limits. The court also required the responsible cabinet secretary to explain the basis for deciding any level of DDT residues to be safe if he proposed continuing to allow residues in food commodities. The ruling in effect endorsed the strict attention to carcinogens that the Delaney clause mandates. By the mid-1970s, "cancer-causing" had become inextricably wedded to the words "toxic chemical" in the popular culture.

With cancer as the ultimate measure of our fears, it was widely assumed that setting levels based on cancer risk would protect humans as well as fish and wildlife from all other hazards as well. So over the past two decades, pesticide manufacturers and federal regulators looked mainly for cancer and obvious hazards such as lethal toxicity and gross birth defects in screening chemicals for safety. Cancer has also dominated the scientific research program exploring possible human health effects from chemical contaminants in the environment. This preoccupation with cancer has blinded us to evidence signaling other dangers. It has thwarted investigation of other risks that may prove equally important not only to the health of individuals but also to the well-being of society.

If this book contains a single prescriptive message, it is this: we must move beyond the cancer paradigm. Until we do, it will be impossible to grapple with the challenges of hormone-disrupting chemicals and the threat they pose to the human prospect. This is not simply an argument for broadening our horizons to recognize additional risks. We need to bring new concepts to our consideration of toxic chemicals. The assumptions about toxicity and disease that have framed our thinking for the past three decades are inappropriate and act as obstacles to understanding a different kind of damage.

Hormone-disrupting chemicals are not classical poisons or typical carcinogens. They play by different rules. They defy the linear logic of current testing protocols built on the assumption that higher doses do more damage. For this reason, contrary to our long-held assumption, screening chemicals for cancer risk has not always protected us from other kinds of harm. Some hormonally active chemicals appear to pose little if any risk of cancer. And as Lindeman and Burlington discovered, such chemicals are typically not poisons in the normal sense. Until we recognize this, we will be looking in the wrong places, asking the wrong questions, and talking at cross-purposes.

Up to now, our concept of injury from toxic chemicals has focused primarily on two things: whether a chemical damages and kills cells as poisons do or whether it attacks the DNA, our genetic blueprint, and permanently alters it by causing a mutation as carcinogens do. With poisoning, the consequences can be illness or death for the affected human or animal. Mutations can eventually give rise to cancer.

At levels typically found in the environment, hormone-disrupting chemicals do not kill cells nor do they attack DNA. Their target is hormones, the chemical messengers that move about constantly within the body's communications network. Hormonally active synthetic chemicals are thugs on the biological information highway that sabotage vital communication. They mug the messengers or impersonate them. They jam signals. They scramble messages. They sow disinformation. They wreak all manner of havoc. Because hormone messages orchestrate many critical aspects of development, from sexual differentiation to brain organization, hormone-disrupting chemicals pose a particular hazard before birth and early in life. As previous chapters have recounted, relatively low levels of contaminants that have no observable impact on adults can have devastating impacts on the unborn. The process that unfolds in the womb and creates a normal, healthy baby depends on getting the right hormone message to the fetus at the right time. The key concept in thinking about this kind of toxic assault is chemical messages. Not poisons, not carcinogens, but chemical messages.

The scientific preoccupation with mapping the human genome and ferreting out the genes responsible for inherited diseases such as cystic fibrosis has created the popular impression that the root of almost all that ails us will be found in our genes. But as must be clear from the scientific work explored in this book, the inherited genetic blueprint is just one factor shaping a baby before birth. Imagine what would happen if somebody disrupted communications during the construction of a large building, so the plumbers did not get the message to install the pipes in half the bathrooms before the carpenters closed the walls. Imagine that the wrong instructions arrived when the program for the climate control system was being set up, and the building thermostat was fixed at eighty-five degrees rather than sixty-eight. Imagine what it would mean if, through a communications mix-up, the high-rise ended up with only one elevator instead of eight.

The construction of a building is as important as the blueprint. A baby's intelligence depends as much on the levels of thyroid hormone reaching the brain during critical periods of development as on inheriting smart genes. A young man may develop testicular cancer because of abnormal hormone levels in his mother's womb rather than because of an inherited cancer gene. As the scientific evidence laid out in this book indicates, synthetic chemicals can obstruct the hormone messages during prenatal development and permanently alter the outcome.

Since hormone-disrupting chemicals do not play by the same rules as classical poisons and carcinogens, efforts to apply conventional toxicological and epidemiological approaches to this problem have typically led to more confusion than enlightenment.

Some critics of EPA regulation have, for example, been arguing that the body can tolerate low levels of contaminants because it has evolved mechanisms that provide a defense against environmental assaults. Operating from the cancer paradigm, they cite the body's ability to repair damaged DNA. As far as we know, however, the body has no analogous repair mechanisms to cope with the hormone-disrupting effects of chemicals. Why? Cells are primed to receive hormone messages, and as we have seen, they readily accept synthetic impostors that mimic natural hormones. The body responds to the impostors as legitimate messengers and allows them to bind to hormone receptors; it does not recognize their action as damage that needs to be repaired.

Hormone systems do not behave according to the classical dose-response model that informs our thinking about biological responses to perturbations. The practice of toxicology and epidemiology rests on the principle first articulated in the sixteenth century by Paracelsus, a Swiss physician who is considered by some to be the father of toxicology. He observed that things that are not poisonous in small quantities can be lethal in larger doses, hence his axiom: the dose makes the poison. Implicit in this axiom is the notion of the classical dose-response curve, where the biological response to a foreign substance increases as the dose becomes greater. This is the operating assumption in epidemiological studies that seek to identify the toxic effects of synthetic chemicals by studying factory workers, who are exposed to higher-than-average levels of contamination.

Such an approach may prove fruitful for toxic substances that act as classical poisons or as carcinogens, though it is also controversial whether the assumption of linearity holds even for carcinogens. In the case of hormonally active chemicals, any study that assumes linearity is bound to yield confusing results because the response does not necessarily continue to increase as doses increase. As described in Chapter 10, high doses may, in fact, produce *less* of an effect than lower doses. Such dose-response curves are not unusual in hormone systems, where the response will increase with escalating doses at first, but then it typically peaks and decreases as the doses climb still higher.

Those studying the endocrine system do not fully understand why this happens. The reason may lie in the basic ways that hormones work through receptors and in the complexity of feedback loops that characterize the endocrine system as a whole. Thus a natural hormone or a chemical impostor can produce effects at low levels because very few receptors are needed to trigger a response. In the case of estrogen, the hormone needs to bind to only one percent of the receptors contained in a cell to stimulate cell proliferation. But as the level of hormones or hormone mimics rises higher and higher, the system eventually responds as if to an overload by shutting down and showing little or no response. With high doses, the cells actually lose receptors, so the cells can no longer respond until hormone levels again drop to low levels long enough for the receptor system to recover.

As EPA toxicologist Linda Birnbaum has noted, most epidemiological studies have focused on adults, typically adult men. This bias is particularly problematic in regard to endocrine-disrupting chemicals. Health studies often look for harmful effects among workers exposed to high levels of toxic chemicals in factories and chemical plants, but these high levels may produce less damage in adults than much lower levels in individuals exposed second hand in the womb. Timing, as we have seen, may be more important than dose, and one may find more telling results by studying the second generation exposed in the womb than by studying those who were exposed only as adults. A major study following the chemical factory accident in Seveso, Italy, focused primarily on the question of whether the high-level dioxin exposure has increased cancer rates among accident victims. Although one study did look for obvious birth defects, this investigation did not consider damage invisible at birth, such as delayed effects on the endocrine, immune, and nervous systems. As evidence has emerged about dioxin's powerful impact on the unborn, researchers are looking again at the Seveso population and exploring other possible consequences of the accident that took place almost two decades ago.

The cancer paradigm also hampers the recognition of the effects of endocrine disruption because it characterizes the threat as disease. Hormone-disrupting chemicals can diminish individuals without making them sick. For this reason, there is an urgent need to look for "impaired function" as well as for disorders that fit the classic notions of disease. For example, having a poor short-term memory or difficulty in paying attention because of exposure to PCBs is very different from having a brain tumor. The former are deficits, not diseases, but they can nevertheless have serious consequences over a lifetime and for a society. They erode human potential and undermine the quality of human life. They undermine the ways in which humans interact with one another and thereby threaten the social order of modern civilization.

Exposure to a hormone-disrupting chemical before birth does not produce just a single clear-cut effect, and in this way, too, the results defy our prevailing notions of chemically induced disease. Depending on the dose and the timing, a foreign chemical can derail development in a variety of ways that will become evident at different times. For example, a boy exposed before birth to chemicals that mimic estrogen may have undescended testicles at birth, a low sperm count at puberty, or testicular cancer in middle age because of this prenatal hormone disruption. These are effects that manifest themselves in many shades of gray rather than in the black-and-white distinctions made between health and illness.

To screen for chemicals that can rob human potential before birth, it will be necessary to look for developmental effects across three generations—those individuals exposed only as adults and their children and grandchildren who inherit hand-me-down poisons. Although this book focuses largely on threats to the middle generation—the first generation to be exposed in the womb—the hormone disruption experienced by these individuals could potentially affect the next generation as well. Those exposed prenatally to endocrine-disrupting chemicals may have abnormal hormone levels as adults, and they could also pass on persistent chemicals they themselves have inherited—both factors that could influence the development of their own children. It will be necessary to invest in more multidisciplinary studies to better diagnose and understand the effects of hormone disruption. Despite alarming signs, such as the report of dropping male sperm count, the lion's share of research money for investigating the effects of environmental contamination of human health still goes into cancer studies. Leading researchers investigating hormone-disrupting chemicals frequently find it impossible to get funding to pursue their work, even though they have already produced landmark studies pointing to profound risks for wildlife and humans.

If we are to come to grips with this threat, we must also shift to a different way of making judgments about environmental contaminants. There is little chance of showing a simple cause-and-effect link between any one or selected groups of hormone-disrupting synthetic chemicals and problems such as the drop in human sperm count that we have already witnessed. Risk assessment in the real world must respond to real problems in real time.

To address this need, some in the environmental field have begun developing an assessment method known as eco-epidemiology. This method, which was pioneered by Glen Fox, Canadian Wildlife Service, and Michael Gilbertson at the International Joint Commission, the advisory body to the United States and Canadian governments on Great Lakes policy, draws together information from a variety of sources, including wildlife data, laboratory studies, and research on the mechanisms of hormone action or toxicity, and makes pragmatic judgments based on the entire body of evidence. In this approach, one assesses the totality of the information in the light of epidemiological criteria for causality, such as whether the exposure precedes the effect, whether there is a consistent association between a contaminant and damage, and whether the association is plausible in light of the current understanding of biological mechanisms. But this real-world environmental detective work comes to judgment based on "the weight of the evidence" rather than on scientific ideals of proof that are more appropriate to controlled laboratory experiments and the practice of science than to problem solving and protecting public health in the real world. As some have noted, it is akin to the decision-making process a physician uses to diagnose a case of appendicitis—where failing to act has grave consequences. In the same way that accumulating evidence and common-sense inferences led to the conclusion that smoking causes lung cancer, it may soon be possible to conclude, if not prove, based on the weight of the evidence, that hormone-disrupting chemicals are linked to testicular cancer, falling sperm counts, and learning disabilities and attention deficits in children.

Cancer is a dramatic disease with devastating effects on victims and their families. It poses little threat, however, to the survival of animal and human populations as a whole. While cancer is tragic on a personal level, healthy populations can quickly replace individuals lost to the disease.

Because hormone-disrupting chemicals act broadly and insidiously to sabotage fertility and development, they can jeopardize the survival of entire species—perhaps in the long run, even humans. This might be hard to imagine in a world facing soaring human numbers, but the sperm count studies suggest environmental contaminants are already having an impact on the human population as a whole, not just on individuals. In their assault on development, these chemicals have the power to erode human potential. In their assault on reproduction, they not only undermine the health and happiness of individuals suffering from infertility, they attack a fragile biological system that over billions of years of evolution has allowed life to miraculously recreate life.

DEFENDING OURSELVES

THE THREAT EXPLORED IN THIS BOOK MAY SEEM OVERWHELMING, ESPE-
cially to those confronting it for the first time. Feelings of fright and
helplessness are, in our experience, not unusual. This is indeed a
frightening problem. No one should underestimate its seriousness,
even though the magnitude of this threat to human health and well-
being is as yet unclear. It would likewise be dangerous to retreat into
denial, which can be a strong temptation in the face of large, insidious problems that leave individuals feeling helpless and hopeless.

But however grim and unsettling the facts appear in this instance, facts are not fate. Trends are not destiny. Three decades ago,
Rachel Carson's predictions about the impacts of synthetic pesticides led to major changes in their use and thus prevented much of
the apocalyptic "silent spring" she envisioned. Today, the growing
scientific knowledge about endocrine-disrupting chemicals gives us
similar power to avert the hazards outlined in previous chapters.
This should be reason for hope rather than despair.

Unfortunately, however, solutions to this problem will be neither quick nor easy. Much of the concern about hormonally active

synthetic chemicals arises from the persistence that many of them have in the environment. Many don't readily degrade into benign components. A generation after industrial countries stopped the production of the most notorious of these persistent chemicals, their legacy endures in food and in human and animal bodies. Some will be in the environment for decades, and in a few cases even centuries. At the same time, other hormonally active chemicals remain in production, and unexpected new sources of exposure continue to come to light. Most disturbing of all, many of us already carry contamination levels that may put us and our children at risk.

Defending ourselves from this hazard requires action on several fronts aimed at eliminating new sources of hormone disruption and minimizing exposure to hormonally active contaminants already abroad in the environment. This will entail scientific research; redesign of chemicals, manufacturing processes, and products by companies; new government policies; and efforts by individuals to protect themselves and their families. Tragically, there is no way to repair the damage done to individuals who now suffer impairments stemming from chemically caused disruption during their early development. Such damage cannot be undone. But with diligent work by government bodies, scientists, corporations, and individuals, we can reduce the threat to the next generation. Over time, the ill effects now evident in wildlife and humans could diminish and gradually disappear.

That is the good news in this troubling picture. Although hormone-disrupting chemicals can cause grievous, permanent damage to those exposed in the womb, they do not attack genes or cause mutations that persist across generations. They have not altered the basic genetic blueprint that underlies our humanity. Remove the disruptors from the mother and the womb, and the chemical messages that guide development can once again arrive unimpeded.

Up to now, women have generally assumed that they could help ensure the health of their children by being vigilant during pregnancy about what they eat and drink and about exposure to X rays, pesticides, and other toxic chemicals. Such short-term prudence will certainly protect the unborn child from many kinds of permanent damage, including the devastating neurological effects of alcohol. But protecting the next generation from hormone disruption will require a much longer vigilance—over years and decades—because the dose reaching the womb depends not only on what the mother takes in during pregnancy but also on the persistent contaminants accumulated in body fat *up to that point in her lifetime*. As discussed earlier, women transfer this chemical store built up over decades to their children during gestation and during breast-feeding.

Thus, it is critical that we as individuals and as a society make choices that reduce this chemical legacy that is being passed from one generation to the next. In the interest of the coming generation and those that follow, we must limit what children are exposed to as they grow up and keep the toxic burden that women accumulate in their lifetimes prior to pregnancy as low as possible. Children have a right to be born chemical-free.

Our day-by-day choices as consumers will have dramatic effects on such exposure and, potentially, an impact that ripples across generations. The food we ourselves eat may help safeguard our children. The way we raise and feed our daughters may help protect our grandchildren.

There are admittedly many unknowns and uncertainties, but until definitive answers are available, a few simple guidelines can help prevent unneeded risk.

Know Your Water

You have a right to know what is in your water. Consider the integrity of your water supply, and don't be lulled into false assumptions about its safety. If you drink well-water, be concerned about groundwater contamination, especially if you live in the Midwest or other agricultural areas. The risk to drinking water supplies may be greatest during and immediately after the peak seasons for pesticide application.

If your water comes from a community source, find out about your water authority's testing program and what it has found. Urge them to test at least monthly and to make the results public. They have a fundamental responsibility to tell you what is in your water and allow you to make judgments about what risks to take. Tell your water officials you are interested in whether they have tested for hormone-disrupting chemicals, especially the herbicides atrazine and dacthal. These chemicals are often sentinels. If either is found, other pesticides are likely to be present as well, and weekly testing is warranted during the growing season, when farmers are applying pesticides to their fields.

Testing your own water is expensive, and few laboratories serving the consumer market are capable of making a thorough assessment of hormone-disrupting chemicals. But a new generation of tests now under development may soon change the practical options available to individual homeowners. Until then, consumers will have to make sure that their public water company is doing sufficient testing to verify the safety of drinking water. Although many hormone-disrupting chemicals are chlorine-containing compounds, the treatment of drinking water with chlorine is unlikely to contribute to the hazards of hormone disruption.

Do not count on filters that are designed primarily to remove bacteria, microorganisms, and unpleasant tastes and odors. They may not remove the hormonally active synthetic chemicals.

Do not assume that bottled springwater is properly regulated or uncontaminated, especially if bottled in plastic.

People living in areas with questionable water supplies may wish to distill their drinking water as they work to improve the quality of their public water supply. Home distilling units are available. But distillation is only a radical, short-term step. It is wholly impractical as a widespread solution to water contamination.

Choose Your Food Intelligently

Clean fish is one of the most healthful sources of animal protein. Yet, as we've seen, fish can also be a source of contamination. For this reason, consumers should scrupulously heed any warnings about fish contamination. Because of concerns about lost license revenues and tourist dollars, public officials are seldom hasty about imposing warnings about contamination in fish and rarely, if ever, do they do so without extremely good cause. In the United States, state fish and game departments are usually the agencies that issue such fish advi-

sories in cooperation with health officials. These public notices typically advise that pregnant women avoid consuming fish caught in
certain areas and that others limit their intake to a recommended
number of fish meals per month.

If anything, these warnings are often insufficiently prudent. Children and women who are not past the age of child-bearing should
avoid fish contaminated with persistent hormone-disrupting chemicals
such as dioxin, PCBs, and DDE. And it is probably wise for everyone
else to forgo them as well. Any fisherman tempted to ignore the warnings should recall the studies cited earlier before delivering his catch to
his family's dining table. Human studies done by the Jacobsons have
reported that children of mothers who ate contaminated Great Lakes
fish show evidence of delayed neurological development and diminished head size at birth. Helen Daly found the offspring of female rats
fed Lake Ontario salmon to be less tolerant of stress, and parallel human studies done by her colleagues showed evidence of reduced stress
tolerance in children of women who ate Lake Ontario fish.

Avoid animal fat as much as possible. As the journey of the
PCB molecule in Chapter 6 demonstrated, many of these chemicals
travel through the food web in fat and become more concentrated as
they move upward to the top predators such as polar bears and humans. In a 1994 report, the U.S. Environmental Protection Agency
found that meats and cheeses are a major source of dioxin exposure
in the United States today. So eating less animal fat—found in foods
such as butter, cheese, lamb, beef, and other meats—will greatly reduce exposure to hormone-disrupting chemicals. Again, it is particularly important that women minimize the consumption of animal fat
from birth until the end of their childbearing years. They bear the next
generation and the responsibility to protect their children from contamination. Moreover, a family diet rich in vegetables, grains, and
fruits has a multigenerational benefit, for it will reduce the risk of
heart disease and cancer for adults and may help protect your children and grandchildren from prenatal hormone disruption.

Buy or raise your own organically grown fruits and vegetables. If
they aren't available at your supermarket or are too expensive, look
to see if your grocer offers produce that has been tested and found to
have "no detectable residue." Ask your grocer if the grocery chain

screens its food for contaminants or buys from suppliers that do. You have the right to know what is in the food you buy. Encourage your grocers to stock and promote organic produce. Give them a copy of this book. Supporting organic agriculture may help safeguard water supplies as well as reduce your family's exposure to pesticide residues.

Minimize contact between plastic and food and avoid heating or microwaving food in plastic containers or with plastic wrap. Use glass or porcelain for microwave cooking. It is entirely possible that some plastics will turn out to be harmless. But with the discovery that hormone-disrupting chemicals leach out of some plastics, caution is warranted, at least until the research is completed or until the sellers of plastic ware can guarantee their products do not release chemicals into food and beverages.

Researchers are only beginning to appreciate the myriad benefits of breast-feeding, which not only aids in mother–baby bonding but also provides infants with important immune protection and a host of substances that enhance development. At the same time, breast-feeding exposes infants to disturbing levels of chemical contaminants, including a number of known hormone disruptors. According to various studies of breast milk contamination, nursing babies take in the highest doses of contaminants they will experience in their entire lives—levels ten to forty times greater than the daily exposure of an adult. It is indeed tragic that breast-feeding is the only efficient way to remove these persistent chemicals from the human body.

We know too little to judge how the undeniable benefits of breast-feeding balance against the risks of transferring hormonally active contaminants. While we have great concern, it is premature to advise women against breast-feeding. Moreover, some studies suggest that the transfer of contaminants in the womb before birth may have a far greater impact than any transfer taking place during nursing. Thus, by the time of breast-feeding much of the potential impact may have occurred. There is a pressing need for research to determine whether the concentrations of hormone-disrupting chemicals in human milk pose enough of a hazard to make breast-feeding inadvisable for some women, perhaps those having their first child later in life. These older women will generally carry a much higher burden of persistent chemicals than first-time mothers who are twenty. Though cow's milk lacks some of the specific benefits of human milk, it contains only one-fifth the concentration of persistent contaminants because cows are shorter-lived animals, vegetarians, and are constantly eliminating contaminants from their body as they are milked daily. We cannot afford to ignore the pressing issue of persistent contaminants when weighing the merits of breast-feeding against alternatives such as bottle feeding with a formula based on cow's milk.

Avoid Unnecessary Uses and Exposure

Wash your hands frequently. Studies show that many synthetic chemicals vaporize and then settle on indoor surfaces—counters, tables, furniture, clothes—where they can be readily picked up by those who touch them. In fact, indoor air experts now sample for contaminants in buildings by wiping surfaces with special equipment. Developing the habit of handwashing, especially in the case of children who often sit or play on the floor, is a simple, effective way to reduce exposure.

Never assume a pesticide is safe. Anything designed to disrupt living organisms—plant or animal—may also prove harmful to humans or other animals in unexpected ways. Recall EPA researcher Earl Gray's discovery that products designed to kill fungus on fruits and vegetables can interfere with the synthesis of steroid hormones in animals and most likely in humans as well. The casual use of pesticides around homes and gardens for frivolous, cosmetic purposes is risky and irresponsible. In the United States, greater quantities of pesticides are applied per acre in the suburbs than on agricultural land, much of it to support the national obsession with green, weed-free lawns. Studies have found higher rates of cancer in children and dogs living in households that use pesticides in the home and garden. The epidemiological studies done to date have not looked for the kinds of functional and developmental problems described in this book.

Make your own lawn pesticide-free and encourage your neighbors to do so. If they persist in their use of pesticides, insist that they post their lawns at the time of treatment. Keep your children and pets away. Organize within your neighborhood to set strict standards for chemical treatment of lawns. Lawn care services sometimes try to reassure uneasy customers by telling them that the pesticides used are "EPA approved." The Environmental Protection Agency has never screened most of the pesticides now on the market for hormone-disrupting activity, and U.S. Environmental Protection Agency registration is no measure of safety. In fact, chemical companies register with the EPA precisely because a product is potentially harmful. Labeling reduces the legal liability of the manufacturer in lawsuits brought by people harmed by using the pesticide. If necessary, stop growing plants or shrubs that require such chemical support to look presentable and replace them with insect- and disease-resistant alternatives. Pesticides should be used only in genuine emergencies.

Don't be blasé about the risks that come along with household pest control, whether you do it yourself or hire a professional exterminator. Following the label won't eliminate the risks to you and your children, but it will reduce them. Use pesticides in your home only if absolutely necessary and, if you do so, follow the label instructions very carefully. It is also important to keep in mind that most pesticides are mixtures of active and "inert" ingredients, and some compounds used as "inerts," such as the nonylphenols and bisphenol-A—are recognized endocrine disruptors. The pesticide labeling law unfortunately does not require manufacturers to list inert ingredients, and legal "trade secrets" provisions allow them to avoid disclosure to consumers. So you cannot tell by looking at a product label whether a pesticide contains an endocrine-disrupting ingredient.

Make a concerted effort to control fleas on pets without insecticides. This is safer for your pet and your family. Moreover, many flea-control products have become increasingly ineffective because heavy use has hastened the evolution of pesticide-resistant superfleas. Frequent grooming, use of a flea comb, and regular baths with a noninsecticidal shampoo can help keep fleas off your dog or cat. You can prevent fleas from gaining a foothold in your house by vacuuming regularly and thoroughly, especially around cracks and baseboards, and by washing your pet's bedding often. Some recommend sprinkling diatomaceous earth, a natural inert product found on the shelf in garden stores, in your pet's favorite areas to discourage fleas.

Find out how your local merchants treat their stores or facilities for pests. It has come to light that some supermarkets in the United States have been fogging their produce with pesticides. Few states provide effective guidance for pesticide application in retail areas or in hotel rooms. Pesticide-free hotel rooms should be a regular option for health-conscious consumers, just as smoke-free rooms now are. Asking for them shows there is a consumer demand.

Be aware that golf courses present a great potential for exposure. By one conservative estimate, based on a report of pesticide use on Long Island golf courses, golf course managers use at least four times more pesticide per acre than farmers do on food crops. Seven of the fifty-two pesticides used on the Long Island golf courses surveyed disrupt the endocrine system and hormones. Other pesticides in use there have been classified as probable or possible carcinogens. Find out what your local course applies and when, so you can play at other times. Keep your hands away from your mouth while golfing, don't chew on tees, and wash your hands after leaving the course. Do not fish on golf courses or downstream of them.

Give babies unpainted, unvarnished toys made of wood or natural fibers. If your young children must have plastic toys, make sure they do not chew on them.

Improving Protection

While individuals can do a great deal to protect themselves, these efforts must be matched by broad government action to eliminate synthetic chemicals that disrupt hormones.

It is beyond the scope of this book to provide a detailed critique of the laws and regulations relevant to this problem. Nonetheless, it is possible to identify several basic principles that can inform future efforts to improve the laws protecting people and ecosystems.

Following the model of the 1987 Montreal Protocol, an international treaty that mandates the phase-out of chlorofluorocarbons and other ozone-depleting chemicals, the United States and other nations should move quickly to implement comprehensive international treaties to halt the use and ecological dispersal of biologically active persistent compounds such as PCBs, DDT, and lindane. While negotiating such international environmental agreements is admittedly challenging, past experience has shown that governments can come together and act in the face of a genuine threat to human welfare. These protocols on persistent hormone-disrupting chemicals should phase out the production and use of these compounds worldwide and provide institutional and financial support for their containment, retrieval, and cleanup.

As a first step, these protocols should require the prior informed consent of countries that are importing chemicals that become persistent contaminants. The exporting business or agency should be required to notify an international monitoring body of each trade and to notify the importing country of the nature of the compounds and the associated risk.

At the same time, individual nations should move to revise domestic laws governing environmental health standards to ensure that they provide protection from chemicals that interfere with hormones. Such revisions should include the following key points:

- *Shift the burden of proof to chemical manufacturers.* Chemical materials continue to be regulated with very inadequate and incomplete information. To a disturbing degree, the current system assumes that chemicals are innocent until proven guilty. This is wrong. The burden of proof should work the opposite way, because the current approach, a presumption of innocence, has time and again made people sick and damaged ecosystems. We are convinced that emerging evidence about hormonally active chemicals should be used to identify those posing the greatest risk and to force them off the market and out of our food and water until studies can prove their impact to be trivial. Every new compound should be subjected to this test before it is allowed to enter into commerce. The tool of risk assessment is now used to keep questionable compounds on the market until they are proven guilty. It should be redefined as a means of keeping untested chemicals off the market and eliminating the most worrisome compounds in an orderly, timely fashion.

■ *Emphasize prevention of exposure.* Many hormone disrupting chemicals alter normal developmental processes, causing permanent consequences that cannot be reversed or even mitigated through later treatment. Because these effects are usually irreversible, treatment after the fact is an unsatisfactory solution. The goal must be to prevent exposure to such chemicals in the first place by eliminating the use and release of hazardous compounds.

■ *Set standards that protect the most vulnerable, namely children and the unborn.* Today's standards have been developed based on the risk of cancer and gross birth defects and calculate these risks for a 150-pound adult male. They do not take into consideration the special vulnerability of children before birth and early in life.

■ *Consider the interactions among compounds, not just the effects of each chemical individually.* Government regulations and toxicity testing methods currently assess each chemical by itself. In the real world, we encounter complex mixtures of chemicals. There is never just one alone. Scientific studies make it clear that chemicals can interact or can act together to produce an effect that none could produce individually. Current laws ignore these additive or interactive effects. Regulating as if chemicals act only individually is as unrealistic as assuming that a batter in a baseball game can only score a run for his team if he hits a home run. In real life and in baseball, the bases may already be loaded and a single could well be enough.

■ *Take account of cumulative exposure from air, water, food, and other sources.* The current legal structure, which includes a number of laws addressing pesticides, food safety, water safety, and air pollution, encourages regulators to focus on one avenue of exposure at a time, such as the contaminant levels in drinking water or pesticide residues on food. This type of approach often fails to consider how the exposure from all the different sources—air, water, food, dust, etc.—adds up. Although exposure from any single source may be tolerable, the total from all sources may be unsafe. For this reason, contaminant levels from any single source must be assessed within the context of total cumulative exposure.

- *Amend trade secrets laws to make it possible for people to protect themselves against undesired exposure while preserving any real need for confidentiality.* Trade secrets laws have been enacted to prevent business competitors from gaining an unfair economic advantage by adopting a company's methods without having borne the cost of product research and development. In practice, these laws are routinely used by manufacturers to deny the public access to information about the composition of their products. Since a skilled chemist can discover what a product contains, we are skeptical that trade secrets laws are keeping such information from business competitors determined to find out. One has to ask who is being kept in the dark by trade secrets provisions, save for consumers, who do not have the money to do the chemical analysis. Until manufacturers provide honest and complete labels for their products, consumers will not have the information they need to protect themselves and their families from hormonally active compounds.

- *Require companies selling products, especially food but also consumer goods and other potential sources of exposure, to monitor their products for contamination.* This should begin in the grocery store. Grocers should be able to tell you, when you want to know, whether your food is contaminant-free. The current testing system, implemented by the Food and Drug Administration, is simply inadequate. It doesn't have the money or the manpower to do the job responsibly. The burden for testing should be shifted to the manufacturer and distributor, with the FDA charged with monitoring to ensure compliance.

- *Broaden the concept of the Toxic Release Inventory.* This powerful right-to-know law, enacted in 1986, now requires companies in the United States to disclose the amount of toxic contaminants that escape from their facilities into the environment in the course of normal operations. As the hazards explored in this book make clear, many hormone-disrupting chemicals enter the environment through "purposeful" release in agricultural pesticides, through detergents, and in plastics. The reporting under the Toxic Release Inventory should include this deliberate release through products as well as inadvertent releases during manufacturing. Companies should, therefore, be required to report the quantity of known endocrine-disrupting compounds incorporated into products sold or transferred from each facility.

- *Require notice and full disclosure when pesticides are used in settings where the public might encounter them.* This would include multifamily dwellings; lawns; places of worship; motels and hotels; places where food is stored, sold, or prepared; and day-care centers, schools, colleges, and other places of learning.
- *Reform health data systems so they provide the information needed to make sound and protective policies.* A lack of crucial data on the national and international level cripples our ability to make timely, intelligent decisions. Our ignorance about trends in many areas of human health is truly appalling. We must undertake a concerted effort to build better records of birth defects and symptoms of impaired function with particular attention to reproductive and neurological disorders. This can be done in ways that protect patient confidentiality while satisfying the health research community's need for better, more comprehensive data. Until this kind of scientific data are available, it will be impossible to determine whether important changes are occurring and to respond appropriately to new hazards.

Research Directions

Changes to laws and regulations must go hand in hand with an ongoing scientific research effort to discover more about the impact of hormone-disrupting chemicals, how they do their damage, and how damage can be avoided. The research should be driven by the need to answer a small number of crucial questions:

- How much are we exposed?
- How is the human body really responding to these chemicals?
- What is the impact on ecosystems?
- When and how should the government act?

FORENSIC RESEARCH

A comprehensive research program is needed to determine the effects of hormonally active synthetic chemicals on human health and well-being. Are we indeed seeing more genital defects, increased infertility, and more children with learning disabilities such as hyperactivity and deficits in attention, as reports suggest? Probing such questions will require a sophisticated integration of epidemiology on human populations, animal studies, and laboratory investigations of how these chemicals act at the cellular and molecular level.

Epidemiological studies, which are never easy, will be particularly difficult in this instance. First, researchers face the lack of an uncontaminated population for comparison. No young person alive today has been born without some in utero exposure to synthetic chemicals that can disrupt development. There are only the less exposed and the more exposed. Then, there is the additional problem of a long lag time between exposure to these chemicals and the emergence of ill effects. If problems become evident only years or decades after birth, reconstructing patterns of exposure will be difficult at best. Epidemiologists may find that the best opportunities for teasing out human health effects are in developing countries, where exposure to agricultural pesticides is generally far greater than in the United States today. Anecdotal reports from some of these countries suggest that hormone-disrupting chemicals may be causing pervasive, transgenerational damage, but the lack of basic health data currently makes it impossible to document these reports.

A systematic assessment of plastics and their possible contamination of food should be a high priority. Over the past thirty years, plastic has become central to our food delivery system, so virtually all our food—from springwater to peanut butter—arrives encased in some form or another of plastic packaging. To what extent and under what circumstances are biologically active compounds leaching from plastics into food and beverages? Is this contamination sufficient to pose a health hazard? Are there safe, inert forms of plastic that do not leach synthetic chemicals when foods are packaged or stored in them?

Recent studies have implicated widely used synthetic compounds such as phthalates, an ingredient in plastics, and alkylphenol polyethoxylates, which are found in plastics, detergents, and many other products, in hormone disruption. We need a better understanding of what happens to such compounds in the environment. How do they break down, and what are the possible consequences for ecosystems of the original compound or of the chemicals created as it undergoes degradation by light, bacteria, and other natural processes?

Serious detailed assessments should be undertaken to consider the role of hormone disruptors in several disturbing ecological trends, especially the dramatic decline and loss of frog populations around the world, the series of epidemics that have hit marine mammals, and other notable biological disruptions. Some classic wildlife crashes should be revisited to ask whether hormone disruptors contributed to the declines or perhaps are impairing recovery. The dramatic ninety percent decline in the waterbird population in Florida's Everglades coincided with profound disruption in the natural water flow but also with the burgeoning use of agricultural chemicals in south Florida. The waterfowl along the major U.S. flyways, which have suffered a decades-long decline, spend their winters in habitats that include farmland and wetlands that receive pesticide-contaminated runoff.

RESEARCH ON BIOLOGICAL MECHANISMS AND EXPOSURE

We need a better understanding of how the undisturbed physiological system works in humans, including the normal levels of hormones and how even natural variations contribute to differences among individuals. There is a pressing need for more information on human exposure to synthetic hormone disruptors. How does the mother's exposure translate into what reaches the fetus, and what does this prenatal exposure mean to that individual's development?

RESEARCH FOR REGULATION AND PREVENTION

A research program must be undertaken to establish standards that protect human and ecological health from hormone-disrupting chemicals. Are there permissable exposures? Do they vary from one compound to another?

Any effort to regulate synthetic chemicals that disrupt hormones will depend on improving our ability to detect hormonally active compounds. What types of screening methods allow for quick, efficient, and cost-effective identification of such chemicals? How quickly can researchers develop them for broad use?

The body does not react to hormone impostors as it does to ordinary poisons, as we noted in Chapter 11. A high dose may in some cases have less impact than a low dose—a phenomenon scientists call a nonmonotonic response. Is this a general phenomenon with hormone systems? If it is, this finding will have profound implications for toxicological testing and regulation. Industry representatives often complain that high-dose testing overestimates the risks at low levels, but such testing might, on the contrary, completely miss damaging effects.

The significance of infant exposure to hormone-disrupting chemicals through breast milk should be a top research priority. Nursing mothers transfer substantial amounts of chemical contaminants to their babies, but how much does this matter? Children born to mothers with contaminated breast milk have already been exposed in the womb. Will the additional exposure through breast milk greatly increase the risk they run? Are there breast-feeding regimens that can lower the transfer rate of contaminants from mother to baby while maintaining the benefits?

Redesigning Manufacture and Use of Chemicals

Hormone-disrupting synthetic chemicals are today an inescapable fact of life. They are in our food and water. They reach us through the air and through consumer goods we bring into our homes. They have spread across the face of the Earth and insinuated themselves into virtually every nook and cranny of the food web. There is no way to recall them. That is the dilemma we face. We can, though, as suggested above, reduce the risks of exposure by personal choice and through government action, but such after-the-fact

remedies are inevitably disruptive, difficult, and incapable of eliminating the problem. Once problematic chemicals are at large, there
is only one option—to manage and cope.

Ultimately, one arrives at the question of how to prevent such
hazards in the first place. How can we enjoy the benefits of synthetic
chemicals without putting ourselves and our children at risk? What
can we do to make sure we don't repeat this kind of mistake in the
future? Traditional regulations and pollution prevention practices
provided are only partial solutions.

To answer the question—how to achieve protection—we must
rethink how we make and use synthetic chemicals. We must redesign the practices, processes, and products that create the problem. Here and there, efforts that move in this direction are already
under way. Two advocates for fundamental rethinking and redesign—Dr. Michael Braungart, a German chemist, and William
McDonough, an American architect—have also been working on a
set of overall criteria to guide such efforts, criteria for the synthetic
chemicals themselves as well as for the processes and products that
contain them. While this movement is still in its infancy, it signals
the direction for changes that will diminish hazards by reducing
waste and the contaminants reaching the environment.

Braungart identifies several guidelines for the production of
chemicals that will make them easier to track and recycle:

- *Greatly reduce the number of chemicals on the market.* With one
  hundred thousand synthetic chemicals in commerce globally
  and one thousand additional new substances coming onto the
  market each year, there is little hope of discovering their fate in
  ecosystems or their harm to humans and other living creatures
  until the damage is done.
- *Reduce the number of chemicals used in a given product; make
  them simpler.*
- *Make and market only chemicals that can be readily detected at
  relevant levels in the real world with current technology* . Some
  compounds currently in broad use are very hard to measure in
  the world at large, making it difficult, economically and practi-
  cally, to study human exposure or their fate in the environment.

- *Restrict production to only products that have a completely defined chemical makeup and stop production of products containing unpredictable mixtures of chemicals.* Such mixtures—for example the 209 PCBs—are difficult to test for safety and to track once released in the environment.
- *Do not produce a chemical unless its degradation in the environment is well understood.* In some instances, chemicals released into the environment can break down into substances that pose a greater hazard than the original chemical.

Braungart and McDonough also advocate a major change in the way we use synthetic chemicals in products and industrial processes, guided by the axiom that there should be "no such thing as waste." This notion borrows from natural systems, where chemicals, nutrients, and organic matter are continuously recycled. The waste from one creature or process becomes resources or food for another. One can see this principle at work in the backyard compost pile where worms, insects, and bacteria transform leaves, grass clippings, carrot peels, and wilted lettuce into crumbly rich black soil to nourish new trees, grass, and vegetables.

Waste from one process could feed another *in the industrial realm as well*, McDonough and Braungart argue. But whether in the compost pile or the factory, such recycling will become possible only if "waste" is not contaminated by substances that make it unusable by living things or for subsequent industrial activities. Through proper design of the manufacturing process and products, McDonough and Braungart believe that most discarded material can "feed" the next process. In a well-designed system, solvents should clean again and again, not just once. Worn-out televisions and other appliances could be returned to the makers, where components would be disassembled and the materials recycled and used again in parts for new televisions.

Although systems designed according to this principle are profoundly different from current approaches, they are not infeasible or impractical, even right now. This concept is already having a profound impact on the automobile industry worldwide. Spurred in part by proposals in Europe to require manufacturers to take back whatever they make, a new trend—products designed for disassembly, or

DFD—is taking off at a gallop. Outside of Detroit, the Big Three auto makers are working jointly to design cars that can be easily taken apart at the end of their life and truly recycled into new automobile parts. Products designed without thought for recycling often contain an assortment of different synthetic materials that make true recycling impossible. Mixed plastics in an auto dashboard, for example, might end up "down-cycled" into a park bench, but they cannot make an encore in a new dashboard. Closing the loop and using materials again and again eliminates the demand for new raw materials and reduces the contaminated waste disposed of in the environment. The key to such closed loop recycling is intelligent design.

In a similar pioneering effort with the textile industry, McDonough and Braungart have helped design a line of upholstery fabric so the manufacturing process and the final product are free of hazardous chemicals. The effort, undertaken with Design Tex, New York–based textile designers and distributors, began with a survey of the 7,500 chemicals used to dye or process fabrics that aimed to eliminate those that pose hazards because they are persistent, mutagenic, carcinogenic, or known to interfere with hormone systems. Only 34 chemicals survived this screening process. The fabric, which is now in production in Switzerland, is a mixture of wool and the plant fiber ramie that comes in a normal range of colors and sells for a price competive with that of comparable fabrics manufactured using conventional methods and design.

Our use of pesticides is also ripe for approaches based on rethinking and redesign rather than the continual use of ever more new chemicals.

■ Over the past forty years, crop losses have remained constant despite greatly increased pesticide use, in large part because of changes in agricultural practices and standards and because of pests' remarkable adaptability. Armed with a chemical arsenal, farmers abandoned common-sense agricultural practices that had been used for millenia to discourage pests, including crop rotation, carefully timed planting, crop diversity, and field sanitation. During this period, farm operations have moved into areas where pest problems had previously made farming infeasible.

- Consumers, food processors, wholesalers, and supermarkets increasingly demand picture-perfect produce that is free of cosmetic blemishes caused by insects, fungus, or disease. Such flaws are not harmful, nor do they make a fruit or vegetable less nutritious, but expectations for picture-perfect produce greatly increase pesticide use. In oranges, for example, sixty to eighty percent of the pesticide use takes place to improve the cosmetic appearance of the skin. It can't be argued, in this or any similar case, that pesticide use is necessary to achieve more or better food.

- Garden designers, garden writers, and homeowners need to take up the challenge of creating a new standard of suburban beauty, one that moves away from the green-carpet aesthetic and revels in a diversity of plants adapted to local conditions. Because this ideal of a homogeneous, flawless greensward fights a natural tendency toward diversity, it is demanding by its very nature, requiring fertilizers and pesticides, frequent watering, and a great deal of time and effort. Like flawless fruit, flawless lawns come at a high price. The time has come to change our attitudes and redesign our yards and gardens with plantings that grow comfortably in the place we live and with mowed play areas that will flourish without constant chemical support. A pioneering team from Yale University recently published a manifesto for this suburban revolution, titled *Redesigning the American Lawn: A Search for Environmental Harmony*, which includes practical guidance for nonspecialists seeking to make their yards safe, saner havens.

The synthetic pesticides developed over the past half century are powerful weapons that should be used sparingly and only when essential. The *other* tragedy of pesticide use—one distinct from the focus of this book—concerns the growing problem of pesticide and antibiotic resistance among insects and disease-causing organisms. Through casual and excessive use of pesticides and drugs, humans have accelerated the evolution of insects, weeds, and bacteria that are increasingly immune to our miracle pesticides and wonder drugs. The bugs are not only fighting back, they are winning this evolutionary struggle. Within a few years of DDT's introduction, new super-bugs had appeared that were invulnerable to its poison. Two decades later, Rachel Carson warned in *Silent Spring* about ever increasing resistance among pests and the ominous implications for human health. Now, at a time when some public health scientists fear an increasing threat of tropical diseases in the United States, the pesticides we need to control disease-carrying insects may no longer be effective. Resistance has become so widespread that we may soon find ourselves as defenseless in the face of disease and health-threatening pests as we were half a century ago. What we thought was a stunning technological conquest of nature is proving only a temporary victory. By using our wonder drugs and miracle pesticides in excess, we have squandered their benefits.

13

LOOMINGS

ANY READER WHO HAS COME WITH US THIS FAR, WHO HAS FOLLOWED the path from the gull colonies on Lake Ontario and the swamps of Florida to university laboratories and then to the doctor's office, must have paused at some point to wonder whether these symptoms have anything to do with the ills of modern human society. Such questions inevitably leap to mind when one learns that gulls in contaminated colonies neglect their nests, or that male mice born to mothers fed with pesticides are much more territorial and potentially aggressive as adults than those who did not have this prenatal exposure. At the moment, there are many provocative questions and few definitive answers, but the potential disruption to individuals and society is so serious that these questions bear exploration.

Declining sperm counts loom ominously over this discussion, for these reports harbor implications that extend beyond the question of male fertility. Animal experiments indicate that contamination levels sufficient to impair sperm production may affect brain development and behavior as well. Thus, it is likely that sperm counts are just one concrete, measurable signal of much broader effects on aspects of human health and well-being that are not so easily quantified. What is at stake is not simply a matter of some individual destinies or impacts on the most sensitive among us but a widespread erosion of human potential over the past half century. The evidence taken as a whole makes it difficult to avoid questions about the significance of this chemical assault for society at large.

Wildlife data, laboratory experiments, the DES experience, and a handful of human studies support the possibility of physical, mental, and behavioral disruption in humans that could affect fertility, learning ability, aggression, and conceivably even parenting and mating behavior. To what extent have scrambled messages contributed to what we see happening around us—the reproductive problems seen among family and friends, the rash of learning problems showing up in our schools, the disintegration of the family and the neglect and abuse of children, and the increasing violence in our society? If hormone-disrupting chemicals undermine the immune system, could they be increasing our vulnerability to disease and, thus, contributing to rising health-care costs? Most fundamentally, what does this mean for the human prospect?

If these effects are occurring broadly, hormone disruption may well be contributing to aberrant and unhealthy tendencies in our society. On the other hand, it is doubtful these chemicals are *causing* all the social dysfunction we see around us. Those seeking a single, simple explanation for such complicated phenomena are bound to be frustrated and disappointed.

Even in the case of relatively straightforward physical problems, such as sperm count declines, we understand far too little about the hormone-disrupting chemicals unleashed in the environment to assess the prospects with confidence. The four studies reported to date show a precipitous drop in human male sperm counts in recent decades—a loss on average of one million sperm per milliliter of semen a year. Such a sharp downward trend is truly alarming. Even more alarming is the fact that this decline continued for almost a half a century before medical researchers recognized what was happening. Will this stunning rate of loss continue? Where will it end?

If currently regulated persistent chemicals are largely responsible for the decline, sperm counts could begin rebounding around

2030. As noted in Chapter 10, the studies find a correlation between the quantity and quality of sperm and the date of a man's birth, with the youngest men showing the lowest sperm counts and the greatest number of malformed sperm—a pattern that strongly supports the theory that the decline is the result of damage before birth or early in life. There is inevitably a long delay, however, before damage becomes evident through a sperm count analysis. The youngest men in the recently reported sperm count studies were born in the early 1970s, just at the time that the United States and other industrialized countries began to restrict the use of highly persistent organochlorine chemicals such as DDT, dieldrin, lindane, and PCBs. So their low sperm counts could reflect the high exposure of their mothers to persistent chemicals in the 1960s and 1970s before governments imposed restrictions. Since then, the concentrations in human tissue of DDT, the DDT breakdown product, DDE, and lindane, for example, have dropped considerably in countries where their use is restricted. If prenatal exposure to endocrine-disrupting pesticides has played a major role in sperm count reductions, one would expect to see an upswing in sperm numbers over the next decade, at least in developed countries, as males born in the 1980s reach maturity. In countries such as India, however, just two persistent pesticides, DDT and lindane, make up at least sixty percent of the pesticides, and their use is still increasing, according to pesticide experts.

Falling sperm counts could be an unfortunate historical episode—an unforeseen consequence of the midcentury experiment with persistent chemicals, which many countries have now wisely discontinued. The threat could now be essentially behind us, even though it may take decades to play out the effects. Unfortunately, the worrisome new discoveries described in previous chapters indicate the hazard from synthetic chemicals has probably not abated. As human exposure to DDT and other persistent compounds has diminished in countries like the United States, exposure to *other* hormone-disrupting chemicals has rapidly increased. Consider the extent to which plastic has replaced glass and paper in packaging over the past two decades. A series of accidental discoveries has demonstrated that plastics are not inert as was commonly assumed and that some of the chemicals leaching from plastics are hormonally active. Plastics have found their way into every corner of our lives, creating the potential for significant chronic exposure to hormone disruptors. They carry everything from soda to cooking oil, they line metal cans and are the preferred material for children's toys. It is unlikely that all plastics are hazardous, but because of manufacturers' claims of trade secrets, there is no way to know the chemical composition of any given plastic container or to judge how much of the plastic in use might be shedding hormone-disrupting chemicals. Scientists also warn that hormone-disrupting chemicals may lurk in ointments, cosmetics, shampoos, and other common products.

It would be comforting to know that hormonally active chemicals are not casting a shadow on the next generation, but the evidence provides no such assurance. As the list of hormone-disrupting chemicals continues to expand, each new addition argues against the likelihood that male sperm count levels will fully recover in the years ahead.

So we find ourselves at an unsettling juncture—uncertain whether the dire trend in human male sperm count will soon bottom out or whether it will continue downward. It is encouraging that some of the most notorious persistent chemicals have been restricted in developing countries and that human body burdens in at least some of these countries have declined as a result. At the same time, surprising discoveries of hormonally active chemicals in unexpected places such as plastics raise new concerns about chronic widespread exposure.

There is always a temptation to extrapolate worrisome trends into apocalyptic, worst-case scenarios, but it is hard to imagine that sperm counts will fall inexorably downward and reach a point that poses an imminent threat to human survival. Even so, humans do appear to be gambling with their ability to reproduce over the long term, which should be of grave concern.

What we fear most immediately is not extinction, but the insidious erosion of the human species. We worry about an invisible loss of human potential. We worry about the power of hormone-disrupting chemicals to undermine and alter the characteristics that make us uniquely human—our behavior, intelligence, and capacity for social organization. The scientific evidence about the impact of hormone disruptors on brain development and behavior may shed new light on some of the troubling trends we are witnessing.

Why did the Scholastic Aptitude Test scores of high school seniors seeking college admission begin to fall sharply from their high point in 1963 and continue downward for almost two decades? Is it solely the result of demographic and social factors, such as changes in the pool of college aspirants or reduced motivation on the part of students, as studies have suggested? What about the problems in our schools? Why can't many children read? Is it because they watch too much TV or spend all their time playing video games, because of a lack of family support for schools, or because they were exposed to PCBs or other thyroid-disrupting chemicals before birth?

While any connection is still speculative, the human and animal studies reporting learning difficulties and hyperactivity in those exposed prenatally to PCBs suggest to us that synthetic chemicals may indeed be increasing the burden on our schools. This seems particularly probable in light of data discussed earlier showing that five percent of the babies in the United States are exposed to sufficient quantities of PCBs in breast milk to affect their neurological development. Moreover, this figure does not take into account the large number of other synthetic chemicals that can also disrupt the thyroid hormones that are vital to brain development. It is difficult to tease this contamination factor out from all the other stresses confronting children in our society—disintegrating families, neglect, abuse, and increasing violence on the streets and even within schools. But save for lead and mercury, educators, physicians, and others have been slow to recognize that the chemical environment may undermine educational efforts as well as the social environment. The hitherto unrecognized hazards of endocrine disruptors need serious investigation, because such disruption could be a major factor in learning and behavioral problems and one that could be reduced in the future through preventive measures.

If such invisible losses are already taking place, they will have greater impact on the society as a whole than on any individuals. Some human studies have suggested that contaminants at levels currently found in the human population could impair mental development enough to cause a five-point loss in measurable IQ. If this happened to a typical child, the consequences would be unfortunate but not catastrophic. Although the child would not fulfill his true potential, he would still fall within the normal range of intelligence and, with discipline, might do well enough in school and gain entrance to a college. But the five-point IQ difference might mean that he lacks the competitive edge to get into a top university.

Consider, however, what it might mean for our society if synthetic chemicals are subtly undermining human intelligence across the entire population in the same manner that they have apparently undermined human male sperm count. With the current average IQ score of 100, a population of 100 million will have 2.3 million intellectually gifted people who score above 130. Though it might not sound like much, if the average were to drop just five points to 95, it would have "staggering" implications, according to Bernard Weiss, a behavioral toxicologist at the University of Rochester who has considered the societal impact of seemingly small losses. Instead of 2.3 million, only 990,000 would score over 130, so this society would have lost more than half of its high-powered minds with the capacity to become the most gifted doctors, scientists, college professors, inventors, or writers. At the same time, this downward shift would result in a greater number of slow learners, with IQ scores around 70, who would require special remedial education, an already costly educational burden, and who may not be able to fill many of the more highly skilled jobs in a technological society. Given the daunting array of problems we face as nations and as a world community, the last thing we can afford is the loss of human intelligence and problem-solving powers.

The animal studies raise even more disturbing questions about the possible impact of synthetic chemicals on behavior, which appears particularly sensitive to disruption by hormonally active contaminants. Researchers find evidence of altered behavior long before they find signs of reduced intelligence or impaired fertility. Recalling Helen Daly's rat studies, the pups born to mothers who ate contaminated fish seemed just as intelligent, healthy, and reproductively fit as the control rats, but they showed great changes in behavior, particularly in their extreme reactions to negative events. Could prenatal exposure to environmental contaminants have similar effects on humans? Could they be reducing our ability to cope with stress as well? The first results from the human studies done by Daly's colleagues show a similar intolerance to stress among the children of women who had eaten contaminated Lake Ontario fish.

Other studies suggest that exposure to synthetic chemicals can make animals more prone to aggression. In studies exposing pregnant mice to relatively low levels of the pesticides DDT and methoxychlor, Frederick vom Saal and his team report a much higher rate of territorial urine marking in their male offspring than in males born to unexposed mothers, a behavior that indicates an increased likelihood of aggression between males. In vom Saal's view, the studies show that hormonally active chemicals may have important effects on social-sexual behaviors. "If animals within a population all show changes in social-sexual behavior, marked disturbance of social structure can occur." Other researchers have fed laboratory rats and mice water containing the same levels of chemical contaminants found in rural Wisconsin wells and discovered that the animals drinking contaminated water showed unpredictable outbursts of aggression. While intriguing, any connection between such studies and the rising violence in American society is, at this point, purely speculative. But without question, these findings point to an urgent need to pursue possible links between chemical contaminants, behavior, and aggression in both animals and humans.

What about the breakdown of the family and frequent reports of child abuse and neglect? If scientists have found evidence of careless parenting in contaminated bird colonies, do these chemicals have any role in similar phenomena among human parents? Reacting to reports of growing neglect and violence against children by their parents, some commentators have ventured that there must be something wrong with these people; some basic instincts seem to be missing. Hormones do not *determine* our behavior, but it is likely that they influence mating and parenting behavior in humans just as they do in other mammals. Recent animal studies have been identifying the biological mechanisms involved in the bonding between mammal mothers and their offspring, and between males and their mates—mechanisms that are dependent on hormones. The effects of contaminants on behavior will vary considerably among species, making it impossible to predict any specific effects on humans. But we are confident that ongoing research will confirm that the hormonal experience of the developing embryo at crucial stages of its development has an impact on adult behavior in humans, affecting the choice of mates, parenting, social behavior, and other significant dimensions of our humanity.

Nevertheless, at the moment it is impossible to know whether hormone-disrupting chemicals are contributing to any of the disturbing social and behavioral problems besetting our society and, if so, how much. Each of these problems is immensely complex and the result of a variety of forces acting together. At the same time, studies with animals are clearly showing that disrupting chemical messages during development can have a lifelong impact on learning ability and behavior. Hormone disruption can increase the tendency toward a certain kind of behavior, such as territoriality, or attenuate normal social behaviors, such as parental vigilance and protectiveness. Given this provocative evidence, we should consider chemical contamination as a factor contributing to the increasing prevalence of dysfunctional behavior in human society as well.

Some might find irony in the prospect that humans in their restless quest for dominance over nature may be inadvertently undermining their own ability to reproduce or to learn and think. They may see poetic justice in the possibility that we have become unwitting guinea pigs in our own vast experiment with synthetic chemicals. But in the end, it is hard to regard such a chemical assault on our children and their potential for a full life as anything but profoundly sad. Chemicals that disrupt hormone messages have the power to rob us of rich possibilities that have been the legacy of our species and, indeed, the essence of our humanity. There may be fates worse than extinction.

14

FLYING BLIND

EVERY CREATURE INEVITABLY ALTERS ITS SURROUNDINGS AS IT SCRAM-
bles to make a living. This is a part of life and has been so since microorganisms first began changing the chemical makeup of the Earth's
atmosphere some two billion years ago.

Humans have been no different. We have hunted game, gathered fruit, cleared forests, drained wetlands, planted fields, dammed
rivers, built cities, soiled streams, constructed factories, and thrust
railroads across desolate plains. But for most of the few million years
that humans have trod upon the planet, our impact has been discrete. We have transformed one valley but not the next; one watershed but not its neighbor; one county but not a continent. The scale
of human changes has always seemed slight when compared with
that of the natural forces shaping the planet.

Today this has all changed. The twentieth century marks a true
watershed in the relationship between humans and the Earth. The
unprecedented and awesome power of science and technology, combined with the sheer number of people living on the planet, have
transformed the scale of our impact from local and regional to

global. With that transformation, we have been altering the fundamental systems that support life. These alterations amount to a great global experiment—with humanity and all life on Earth as the unwitting subjects.

Synthetic chemicals have been a major force in these alterations. Through the creation and release of billions of pounds of man-made chemicals over the past half century, we have been making broadscale changes to the Earth's atmosphere and even in the chemistry of our own bodies. Now, for example, with the stunning hole in the Earth's protective ozone layer and, it appears, the dramatic decline in human sperm counts, the results of this experiment are hitting home. From any perspective, these are two huge signals of trouble. The systems undermined are among those that make life possible. The magnitude of the damage that has already occurred should leave any thoughtful, person profoundly shaken.

It is equally disturbing that the global scale of the experiment makes it extremely difficult to assess the effects. Over the past fifty years, synthetic chemicals have become so pervasive in the environment and in our bodies that it is no longer possible to define a normal, unaltered human physiology. There is no clean, uncontaminated place, nor any human being who hasn't acquired a considerable load of persistent hormone-disrupting chemicals. In this experiment, we are all guinea pigs and, to make matters worse, we have no controls to help us understand what these chemicals are doing. Faced with the question of whether synthetic chemicals are contributing, for example, to learning disabilities, researchers have typically set up studies comparing contaminated children with an uncontaminated control group. Tragically, no children today are born chemical-free. In the search for relatively uncontaminated control populations, researchers have ironically discovered the appalling universality of this contamination. Even Inuits living a traditional lifestyle in remote regions of the Arctic have not escaped. The pollution has come to them.

The early results from this unintended experiment raise thorny and profound questions that reach far beyond the immediate challenge of managing and eliminating the chemicals that have caused

these problems. It is no longer sufficient to look for the next round
of substitutes for existing chemicals, for a new generation of suppos-
edly less damaging synthetic compounds. The time has come to shift
the discussion to the global experiment itself.

What has this breathtaking plunge toward new technologies
wrought? It has yielded unparalleled health, luxury, and comfort for
some significant minority, at least, of the human population, but the
technologies themselves have often had a dark side that has only become evident decades later, after it is too late to recall them. When
questioned about the risks of releasing genetically engineered organisms into the environment, one of the world's preeminent molecular
biologists saw no reason for hesitating. He told a group of journalists
that our society has to "be brave" and forge ahead with new technologies despite the uncertainties. But what seems brave to some
seems foolish to others.

If the ozone hole and falling sperm counts are clear warnings
about the perils of proceeding with business as usual, where do we
go from here? Is there any way to anticipate the consequences of
our technology? If we remove hormone-disrupting chemicals from
the market, how can we be sure that their replacements won't be creating other nasty surprises thirty years hence? Is there any way to
stop the experiment with our children and the environment, an experiment that has been an accepted way of life in the twentieth century? Or is the prospect of such hair-raising surprises a part of the
Faustian bargain we have made in exchange for health, comfort, and
convenience?

When stopped short by one of these nasty surprises, such as the
ozone hole, we have typically set about in search of "safe" substitutes—a quest based on the unarticulated assumption that synthetic
chemicals can be put into the environment with impunity if chemical companies and government regulators screen them properly for
safety. But proposed substitutes that may be "safe" for the ozone
layer pose other hazards, it turns out, through their capacity to trap
heat and accelerate greenhouse warming.

A similar pattern emerges in the history of pesticide oversight.
Like generals, pesticide regulators are always and perhaps inevitably fighting the last war. Again and again, they have vetted chemicals

for the most recently recognized hazard only to be blindsided by dangers they never thought to anticipate. They judged DDT by the hazards of the previous generation of pesticides—the acutely toxic arsenic compounds that could bring sudden death to farmers or those unfortunate enough to eat food contaminated with residues. Only after DDT had been spread as liberally as talcum powder across the face of the Earth did we realize that DDT brought death as well, but in a different way. When concerns emerged about the persistence of DDT and its impact on wildlife, regulators imposed controls, and less persistent compounds such as methoxychlor came onto the market. Now we know that methoxychlor, which is still in wide use, disrupts hormones.

There is a need to screen the thousands of chemicals in commerce and to eliminate those that disrupt hormones. If we proceed, however, as we have in the past, we will simply spread a new generation of substitute chemicals across the face of the Earth. It will be yet another chapter in this reckless experiment. Though these new chemicals may be safer from the perspective of hormone disruption, it is likely they will have other unforeseen consequences—some relatively trivial and some perhaps as serious as the ozone hole.

Judging from past experience, it may take a generation for the next nasty surprise to emerge. When it comes it will show up where we least expect it. Thirty years from now, our children may be struggling to stem another serious assault on the systems that support life. Perhaps the next surprise will show up in the soil, one of the least appreciated parts of our life-support system. The consequences would be dire indeed if human activities were to seriously undermine the soil's ability to recycle nutrients—a process of recycling and renewal that depends on a myriad of bacteria, fungi, and insects. The safer bet, however, is that the surprise will be something never even considered. If anything is certain, it is that we will be blindsided again.

This caution does not arise from any propensity for pessimism or dislike of technology. It arises from the very nature of our global experiment and from our inescapable ignorance, which makes it impossible to foresee consequences or guarantee safety. The dilemma is simply stated: the Earth did not come with a blueprint or an instruction book. When we conduct experiments on a global scale by releasing billions of pounds of synthetic chemicals, we are tinkering with immensely complex systems that we will never fully comprehend. If there is a lesson in the ozone hole and our experience with hormone-disrupting chemicals, it is this: as we speed toward the future, we are flying blind.

We can screen chemicals for hazards that have already confronted us such as hormone disruption and ozone depletion, but the next nasty surprise will happen because we do not even know what questions to ask. Nothing illustrates this point better than our experience with two now infamous chemicals—CFCs and DDT.

Like DDT, ozone-depleting chlorofluorocarbons (or CFCs) were touted as one of the safest substances ever invented, and like DDT, they seemed one of the unalloyed blessings of progress when they were first synthesized by Thomas Midgley Jr. in 1928. Midgley, one of the pioneers in industrial invention, developed CFCs in response to the demand for a safer alternative to the toxic and flammable chemicals used as coolants in refrigerators. In 1941, he received chemistry's highest award for his work, the Priestley Prize. Making his acceptance, Midgley, a man with an incorrigible theatrical streak who loved to play Mr. Wizard, could not pass up the opportunity to treat the audience to one of his favorite demonstrations. He poured CFCs into a shallow dish, inhaled as the refrigerant vaporized, and held his breath as he lighted a candle. Then he exhaled, extinguishing the candle triumphantly—again demonstrating that the chemical was neither flammable nor toxic to humans and, therefore, unquestionably safe.

CFCs were on the market for more than forty years before the first shadow of suspicion fell on them. In 1970, James Lovelock, the maverick scientist and inventor who would later become widely known for the Gaia hypothesis, began to make measurements of the atmosphere with his new invention—an electron capture detector that increased the sensitivity of the gas chromatograph a thousandfold. With this powerful new tool, it was now possible to detect minute traces of synthetic chemicals in the atmosphere that are present in concentrations of parts per trillion. Lovelock soon began to find CFCs everywhere he looked, even in samples taken from a ship that cruised to the southern tip of South America—a sign that CFCs were now ubiquitous in Earth's atmosphere.

By 1972, Lovelock had communicated his findings to Raymond McCarthy, an official at Du Pont, the world's leading manufacturer of CFCs. Apparently concerned by the news that CFCs were accumulating in Earth's atmosphere, McCarthy called a meeting of the CFC manufacturers in the chemical industry to discuss "the ecology of CFCs in the environment." At the same time, he commissioned several studies to consider the reactivity of CFCs.

Du Pont was no doubt reassured by the findings from these studies, which concluded that CFCs did not appear to break down into toxic or reactive compounds that might harm people or cause environmental problems. Unfortunately, however, the eyes of the researchers were fixed only on the lower atmosphere. It appears no one even considered possible threats to the ozone layer high up in the stratosphere. That question would surface two years later in June 1974, when chemists Sherwood Rowland and Mario Molina published their now famous paper in the journal *Nature*, describing how CFCs would eventually make their way to the stratosphere and attack ozone. Ultimately Du Pont would phase out its manufacturing and distribution of CFCs. And in 1995 Rowland and Molina were awarded the Nobel Prize for this research.

The history of DDT contains a similar paradox. The pesticide was considered such a milestone on the road of human progress that its developer, Paul Müller, was hailed as a savior and awarded the Nobel Prize in 1948. In the short term, the chemical did seem wondrous. It killed insects while posing little direct threat to humans, and by eliminating the mosquitoes that carry malaria, it saved countless lives. But like CFCs, DDT was at the same time invisibly attacking the foundation of life.

In the end, what we did not know proved to be more important than what we did know. In the end, what we thought were the safest chemicals proved to be among the most dangerous. And when the ozone depletion predicted by Rowland and Molina appeared, it far exceeded the worst-case scenario that atmospheric scientists had forecast.

* * *

The situation confronting us is not one that lends itself to easy prescriptions or simple answers. Our current economy and civilization are built on a foundation of fossil fuels and synthetic chemicals. According to one chemical industry estimate, chlorinated synthetic chemicals and the products made from them constitute forty-five percent of the world's GNP. If it has taken fifty years to work our way into this dilemma, it will almost certainly take just as long or longer to find our way out of it.

As we look toward the future and think about charting a new course, it is critical to begin with a clear-eyed view of our situation. As the experience over the past half century has demonstrated, there is no way to put large quantities of man-made chemicals into the environment without exposing our children and ourselves to unknown risks. Many of these synthetic compounds may prove harmless, but others may not. We must face the fact that there is no way to guarantee the safety of synthetic chemicals, even those that have been on the market for decades. CFCs had been in broad use for fifty years before the ozone hole was discovered over Antarctica. The lag time before effects emerge in vast, complex systems can give a false sense of safety and increase the opportunity for catastrophe.

We must be ever mindful that for all the advances in science, we still have only the most general understanding of the life systems on which we have been experimenting—whether our own bodies or Earth's atmosphere. At the time that CFCs were invented, scientists did not understand the ozone layer or its importance in shielding the Earth from ultraviolet radiation. That came three years later through the work of a British scientist, Sydney Chapman. DDT and other hormone-disrupting chemicals were on the market for two decades before researchers began to fathom the mysteries of the hormone receptors and even longer before they discovered that synthetic chemicals could mimic hormones and engage those receptors.

Ultimately, the risks that confront us stem from this gap between our technological prowess and our understanding of the systems that support life. We design new technologies at a dizzying pace and deploy them on an unprecedented scale around the world long before we can begin to fathom their possible impact on the global system or ourselves. We have plunged boldly ahead, never acknowledging the dangerous ignorance at the heart of the enterprise.

Such arrogant presumption may be an ineradicable part of human nature. The ancient Greeks called it hubris. Throughout human history, humans have risked the unknown, courting both success and catastrophe. What differs now is the stakes, the magnitude of possible mistakes. Our activities no longer involve just one village and its neighbor, one valley or the next. The scale of human activity means that these experiments engage the planet.

As we race toward the future, we must never forget the fundamental reality of our situation: we are flying blind. Our dilemma is like that of a plane hurtling through the fog without a map or instruments. Instead of being able to provide a reliable radar system, scientists are peering through the cockpit window trying to warn of any obstacles ahead. And usually, the most they can say is that the dark mass looming into view might be a cloud bank. Or then again, it might be a mountain.

So what do we do? Land the plane as quickly as possible, slow down, or proceed full speed ahead because it would be incredibly expensive and disruptive to cancel this trip?

These kinds of questions confront us today as we grapple with the consequences of our half-century experiment with synthetic chemicals. When confronted with a troubling environmental problem, the first impulse has been to appoint a panel of experts in the hopes that they can give us the right answer. Scientists can certainly provide invaluable guidance, as the work described in this book amply demonstrates. But science alone does not always have the answer.

Deciding on a wise course involves a host of considerations and, most of all, value judgments. It is not just a question of the quality of science describing the problem but also of how we see the risks and how much risk we are willing to entertain. Consider the convenience that endocrine-disrupting plastics bring to human lives against the risks they entail. If all that is at stake is the survival of a single gull colony, it may be wise to wait for further scientific study before embarking on an effort to reduce exposure. If, on the other hand, it is a question of decreasing human sperm counts, prudence may dictate acting immediately rather than waiting to see whether the downward trend continues.

Phasing out hormone-disrupting chemicals should be just the first step, in our view. We must then move to slow down the larger experiment with synthetic chemicals. This means first curtailing the introduction of thousands of new synthetic chemicals each year. It also means reducing the use of pesticides as much as possible, for these compounds are biologically active by design, and billions of pounds are deliberately released into the environment each year.

But these steps merely deal with the problems of which we have some inkling, however crude. They help not at all with the next generation of surprises, the next unexpected results from our massive alterations of the planetary system. In this light, eroding ozone and falling sperm counts cast dark shadows across the human future. They confront us with the unavoidable question of whether to stop manufacturing and releasing synthetic chemicals altogether. There is no glib answer, no pat recommendation to offer. The time has come, however, to pause and finally ask the ethical questions that have been overlooked in the headlong rush of the twentieth century. Is it right to change Earth's atmosphere? Is it right to alter the chemical environment in the womb for every unborn child?

It is imperative that humans as a global community give serious consideration to this question and begin a broad discussion that reaches far beyond the usual participants—the chemical companies, government regulators, farmers, economists, scientists, and environmental groups. This discussion must engage teachers and parents, physicians and philosophers, artists and historians, spiritual leaders such as the Pope and the Dalai Lama, and others who reflect the richness and diversity of human experience and wisdom.

On a more practical front, we need to explore whether it is possible to discontinue the global experiment without abandoning synthetic chemicals. Are there principles of chemical design and use that would allow us the benefit of innovative materials without undue exposure and risk? Considering the skyrocketing human population and a daunting agenda of global environmental problems, it seems impossible to turn the clock back half a century and return to a material horizon bounded by wood, steel, and glass. At the same time, any such exploration must always bear in mind that it is impossible to anticipate nasty surprises. The goal must therefore be to keep human and environmental exposure to an absolute minimum. How synthetic chemicals fit into a sustainable, healthy future remains unclear.

If it is too early to describe the precise road, it is possible to signal the direction for this journey. For the past half century, the commerce in cheap, abundant synthetic chemicals has shaped agriculture, industrial processes, economies, and societies. It is impossible to imagine the great migration of Americans to the steamy Sun Belt without the CFCs that made it possible to air-condition homes, cars, and public buildings. Similarly, the new generation of synthetic pesticides that swept onto the market after World War II aided and abetted the growth of specialized industrial farming that depended exclusively on a chemical arsenal for pest control and abandoned agricultural practices such as crop rotation, carefully timed planting, or other methods to keep insects in check. The chemical age has created products, institutions, and cultural attitudes that require synthetic chemicals to sustain them.

The journey to a different future must begin by defining the problem differently than we have until now. As a general rule, the framing of a problem limits solutions more than a lack of ingenuity or technology. The task is not to find substitutes for chemicals that disrupt hormones, attack the ozone layer, or cause still undiscovered problems, though it may be necessary to use replacements as a temporary measure. The task that confronts us over the next half century is one of redesign. When forced by the phaseout of CFCs to reconsider the use of solvents when manufacturing electronic circuitry, one research effort in the United States found a way to eliminate the need for CFCs or any other solvent by redesigning the soldering process. Following such examples, we need to redesign not only lawns, food packaging, and detergents, but also agriculture, industry, and other institutional arrangements spawned by the chemical age. We have to find better, safer, more clever ways to meet basic human needs and, where possible, human desires. This is the only way to opt out of the experiment.

As we work to create a future where children can be born free of chemical contamination, our scientific knowledge and technological expertise will be crucial. Nothing, however, will be more important to human well-being and survival than the wisdom to appreciate that however great our knowledge, our ignorance is also vast. In this ignorance we have taken huge risks and inadvertently gambled with survival. Now that we know better, we must have the courage to be cautious, for the stakes are very high. We owe that much, and more, to our children.

EPILOGUE FOR PAPERBACK EDITION

THE SCIENTIFIC DETECTIVE STORY RECOUNTED IN OUR STOLEN FUTURE continues to unfold at an accelerating pace. With its increasing prominence, the endocrine-disruption hypothesis has inspired a host of studies, reports, and scientific assessment efforts exploring diverse aspects of this complex question. New scientific papers are appearing monthly, and in the time since this book was completed, a number have made important contributions to the scientific case. These studies have answered some questions and raised new ones, but in their collective weight they support concerns that relatively low levels of baseline contamination experienced by broad segments of the human population can undermine prenatal development and harm children.

Nothing in the emerging evidence to date has fundamentally altered or challenged the argument put forward in these pages. As scientists give closer scrutiny to available human health data, they are detecting more signs of damage in children that parallel effects reported in lab experiments and in wildlife. This recent work has emphasized the vulnerability of the brain and the immune system, which appear at least as sensitive as the reproductive system to prenatal disruption from contaminants, if not more so.

In May 1996, a group of international experts warned that endocrine-disrupting chemicals—at levels found in the environment and in humans—threaten brain development and called for an international research effort to investigate this hazard. This warning emerged from a scientific assessment effort in Erice, Sicily, involving eighteen prominent scientists from a variety of disciplines relating to the nervous system, brain development, and behavior. In a consensus statement modeled on the 1991 Wingspread Statement, the participating scientists stressed the extreme sensitivity of the developing brain to chemical disruption and the danger of permanent damage in children that can show up as reduced intelligence, learning disabilities, attention-deficit problems, and an intolerance to stress.

The group, which included scientists from federal agencies as well as academic researchers, expressed additional concerns about the broader societal implications of such disruption. "Widespread loss of this nature can change the character of human societies or destabilize wildlife populations. Because profound economic and social consequences emerge from small shifts in functional potential at the population level, it is imperative to monitor levels of contaminants in humans, animals, and the environment that are associated with disruption of the nervous and endocrine systems and reduce their production and release."

In the view of these experts, the list of chemicals that can disrupt the chemical messages vital to normal brain and nervous-system development is incomplete, but it includes such known disruptors as dioxins, PCBs, some plasticizers, and many pesticides. Because dioxins and PCBs impair normal thyroid function, which plays a key role in brain development, the group stated their suspicion that these contaminants "contribute to learning disabilities, including attention-deficit hyperactivity disorder and perhaps other neurological abnormalities."

Four months later in September 1996, a new paper in the *New England Journal of Medicine* added further urgency to this warning, for it documented lasting intellectual impairment in children exposed to PCBs in the womb. More than a decade ago, Joseph and Sandra Jacobson began a study looking at whether a mother's consumption of Lake Michigan fish, which contain elevated levels of PCBs and other man-made contaminants, had any effect on her children. As we discussed in Chapter 9, they did find measurable losses in motor coordination, short-term memory, and verbal skills as they tracked the development of the children born to women who had eaten two or three fish meals a month in the six years prior to pregnancy. Most recently, when 212 children in the study reached eleven years of age, the Jacobsons gave them a battery of IQ and achievement tests and compared the results with the levels of PCBs found in the mother's blood and milk at the time of the child's birth. The results published in the *New England Journal* show evidence of persistent, measurable intellectual impairment, which has not been overcome by environment or education. Those exposed to higher levels of PCBs showed "deficits in general intellectual ability, short-term and long-term memory, and focused and sustained attention." The most highly exposed group had a loss of 6.2 points in IQ. They were three times as likely to have low average IQ scores and twice as likely to be at least two years behind in reading comprehension.

As the Jacobsons note, the levels of PCBs carried by these women "were similar to or slightly above the general population level in the United States." They also stress that the implications of their findings are not limited to children of women who eat fish from Lake Michigan. "Women who eat no fish may accumulate these compounds from other food sources, including dairy products, such as cheese and butter, and fatty meats, particularly beef and pork."

Another mother-child study in the Netherlands has further emphasized that typical background levels of PCBs in women are having a small but measurable negative effect on children's development. In this investigation the researchers compared a child's scores on psychomotor development tests with PCB levels in its mother's blood during the last month of pregnancy, and found that the children born to women with higher PCB levels lagged behind in tests given at three months of age.

Dutch pediatric researchers also conducted an exploratory study looking for another form of invisible damage—immune system changes linked to PCB and dioxin exposure before and around birth. Here again, the study focused on children born to women with typical background levels of these contaminants. The researchers found small aberrations in the infants' developing immune system which, although not sufficiently large to cause more illness in infancy, may persist and presage later difficulties, such as immune suppression, allergy, and autoimmune disease.

Over the past year, two separate assessment efforts have also given voice to growing concern about immune system hazards that receive little attention in chemical testing and regulation. A consensus statement published in August 1996 from a scientific meeting on the chemical threats to the immune system held at Wingspread concluded: "The potential exists for widespread immunotoxicity in humans and wildlife species because of the worldwide lack of appropriate protective standards. . . . The risk of exposure to known immunomodulators is sufficient to warrant regulatory approaches that would limit exposure." In a March 1996 report from the World Resources Institute, Robert Repetto and Sanjay Baliga review numerous studies showing that many pesticides damage the immune system, concluding "existing evidence of a significant worldwide public health risk justifies both greater efforts to reduce pesticide exposures and much expanded research into pesticide-induced immunosuppression and its health consequences."

No aspect of the extensive scientific evidence considered in *Our Stolen Future* commanded more media attention than the controversial reports over the past four years of declining sperm counts. The question of sperm-count trends is troubling and compelling, indeed, but the intense focus on this single issue has unfortunately obscured the extensive body of scientific evidence gathered in this book and the larger concern about diverse hazards stemming from altered hormone levels during early development. Reduced sperm counts are just one kind of reproductive effect documented in laboratory animal studies and one possible hazard to humans. Moreover, the reproductive system is one target among many for developmental disruption. Within this broader picture, reports of declining human sperm counts are just one element in a complex and cumulative pattern of evidence that supports the endocrine-disruption hypothesis.

Not surprisingly, with the publication of additional studies, the question of sperm-count trends has taken on more complexity. In the time since this book was finished, another major study reporting declining sperm counts has appeared, this one from Scottish researchers, while two U.S. studies reported that sperm counts in four cities on the western side of the Atlantic appeared to be holding steady.

Published in the *British Medical Journal* in February 1996, the Scottish study reported a strong decline in sperm counts over a twenty-year period. It also found that the loss follows the same striking pattern reported in three earlier studies—men born later in the century have lower sperm counts and their sperm is of poorer quality. Scottish men born in the 1970s had 24 percent fewer motile (actively swimming) sperm in their semen samples than men born in the 1950s. Over this same period, sperm concentrations dropped at a rate of 2.1 percent a year with a later year of birth.

In contrast, the two U.S. studies found no sperm-count declines in New York, Minneapolis, Los Angeles, and Seattle. They did reveal striking differences among these cities, however, with New York levels more than twice those in Los Angeles'.

Taken together, these new studies indicate that geographical variation is an important part of this story. Why do men in Los Angeles have less than half the sperm of men in New York? Why are sperm numbers dropping in Edinburgh, Paris, and Brussels but not in Minneapolis or rural Finland?

Unfortunately, these and related questions got poor coverage in many media accounts of the great sperm-count debate. For some reason U.S. news accounts implied that because sperm counts were reportedly not declining in the few U.S. cities studied, there is no problem and no reason for concern—a conclusion that ignores the extensive European data. In fact, the results from the U.S. cities do not contradict the European research because they are not contrary findings about men living in the places cited in the European studies. At most, they show an inconsistency in sperm trends in different places. Human exposure to contaminants differs considerably from one place to another, depending on air currents, water supply, proximity to farming regions or industrial centers, and a host of other factors. Thus one would expect some geographical variation if synthetic chemicals are affecting sperm counts. Indeed, the geographic patterns may provide clues for epidemiologists working to identify why sperm count appears to be declining in some places and not in others. Danish reproductive researcher Niels Skakkebaek noted that testicular cancer rates and genital defects were on the rise as sperm counts were falling, leading him to theorize that all might stem from prenatal hormone disruption. Viewed along with the sperm-count results, these patterns may yield important clues about what is responsible for pronounced regional differences and for the worrisome downward trends that have been reported.

The press coverage also ignored important methodological differences among the studies, failing to note a fundamental weakness of the U.S. studies in New York, Minneapolis, and Los Angeles. This study was based on men who had volunteered for vasectomies—a group known from other studies to show higher than average sperm counts and to be unrepresentative of the population at large. The Scottish study, in contrast, avoided this bias by recruiting men broadly from the general population.

Tackling the sperm decline question in a different way, a recent laboratory study on rats provides new factual support for links between prenatal exposure to estrogenic compounds and adult sperm count. In this experiment a team of British researchers, which included such well-known figures in this field as Richard Sharpe and John Sumpter, gave pregnant rats water containing the synthetic estrogen DES and two common estrogenic chemicals found in the environment—octylphenol, a detergent breakdown product, and butyl benzyl phthalate, a plasticizer that is commonly found in the environment. Exploring possible relevance to humans, the researchers administered low doses—doses they said were roughly comparable to estimated human intake—to the rats throughout their pregnancy and for the twenty-one days they were nursing their pups. Octylphenol and butyl benzyl phthalate produced effects similar to DES, and males exposed to these estrogenic chemicals in early life had smaller testicles and reduced sperm production of as much as 21 percent when they reached adulthood. This is an important finding, but it hasn't yet been confirmed in a repetition of the experiment.

Human health studies are also heightening concern that the contaminant levels encountered in the environment are sufficient to do damage. A study in Minnesota found several strong signs of a link between pesticide exposure and birth defects in that state's farming regions and specifically implicated several endocrine-disrupting compounds including the herbicide 2,4-D in higher rates of anomalies such as urogenital defects. According to the data gathered and analyzed by the team headed by Vincent Garry of the University of Minnesota Laboratory of Environmental Medicine and Pathology, higher rates of birth defects occur not only in the children of farmers, who apply pesticides and are directly exposed to the chemicals at work, but also in the children of families living in predominantly agricultural regions in the state where chlorophenoxy fungicides and herbicides are most heavily used. Children conceived in the spring, when herbicides are routinely applied, are at particular risk of birth defects, the study found. In another intriguing finding, the researchers noted that birth defects occur at a much higher rate in the male offspring of those who apply pesticides than in the general population. One of the most remarkable aspects of this study is the high frequency of the birth defects it reports. With cancer effects, a frequency of one per million prompts regulation. Garry's study shows that birth defects in the children of pesticide applicators in heavy-use areas increased by a rate of at least an additional one per hundred—a rate ten thousand times higher than the federal threshold for cancer effects.

Alarming deformities are also showing up in frogs all across Minnesota, as well as in Wisconsin, South Dakota, California, Vermont, Kansas, Missouri, and the St. Lawrence River Valley in Quebec. Since the first report in August 1995, researchers in Minnesota have found grotesquely deformed frogs at more than one hundred sites. One animal had four legs sprouting from its stomach; another a leg growing out of its neck. The most bizarre, perhaps, was a frog with an eye inside its mouth. The animals appear victims of something that severely derailed normal development. While scientists are exploring a number of theories including disruption caused by parasites or by bacterial or viral disease, pesticides and synthetic chemicals remain leading suspects. As we discussed in Chapter 9, frogs by nature of their aquatic life and permeable skin are at particular risk.

New research has also explored the contentious arguments—raised by industry-funded advocates and scientists after the book's publication—that the impact of *synthetic* hormone disruptors is trivial compared to the far greater quantities of estrogenic compounds humans encounter in foods such as soybeans, sunflower seeds, and alfalfa sprouts. The importance of plant estrogens in total human exposure to hormone-disrupting substances is a legitimate question, which we explored in some detail in Chapter 5. But facile comparisons of tofu and DDT do not do justice to the complex science underlying this issue, or even to the relatively simple observation that man-made compounds such as DDT, dioxin, and PCBs are far more persistent in the human body than plant estrogens.

A new study from researchers at Tulane University and the University of Florida provides added evidence that some natural and synthetic estrogens differ in another extremely critical respect—their ability to circumvent the body's natural defense system. In a pregnant woman, the binding action between special blood proteins and estrogens ties up most of the hormone circulating in the blood and thus acts to protect fetuses from excessive estrogen exposure in the womb. As we noted in Chapter 5, some researchers have suspected that humans acquired some defense against plant estrogens through their long evolutionary experience with such substances in food. At the same time, studies on DES showed that these protective proteins do not bind with this synthetic estrogen, raising fears that the body might be defenseless against other man-made estrogenic compounds as well.

Using a genetically engineered screening method involving human estrogen receptors implanted in yeast cells, the researchers headed by John McLachlan of Tulane and Louis Guillette of the University of Florida probed the important question of whether the body's natural defenses also recognize plant estrogens and man-made estrogens. A decreased estrogen response would indicate that the protective proteins were binding with the estrogenic compounds and inactivating them.

McLachlan's team found that the blood proteins do provide protection from the plant estrogens they tested but do not appear to protect against the man-made estrogens that were used in the experiment. Commenting on these specific results, the team noted that "phytoestrogens bind ... with a higher affinity than synthetic estrogens, suggesting that humans have evolved a mechanism to protect themselves from continuous exposure to phytoestrogens. On the other hand, the relative inability of synthetic estrogens to bind ... indicates a potential fundamental difference in the interaction of estrogens from diverse sources with these proteins."

These experiments by no means resolved the general question of whether evolutionary experience provides protection from *all* plant estrogens. Reproductive impacts from fetal exposure to some plant compounds are well-known, as discussed in Chapter 5, so it seems unlikely that it will come down to a simple case that all plant estrogens are blocked and all synthetic estrogens are not. Continuing research into this issue will no doubt reveal a far more complex picture concerning natural defenses and hazards. Indeed, a key part of the ongoing research will explore why blood proteins are effective against some but not all compounds. While far from the final word, these new results do, however, counter the claim that the presence of plant estrogens in the human diet is sufficient reason to dismiss concern about synthetic compounds. It is clearly not.

With a paper published in *Science* in June 1996, McLachlan and

his team also highlighted the possibility of wildly unpredictable interactions *among* synthetic chemicals. As noted in Chapter 8, solid scientific evidence exists that small, seemingly insignificant quantities of
individual estrogenic chemicals can act together producing additive
effects. Scientists have also been reporting various types of synergy
where a mixture of two chemicals is far more potent than the individual
compounds.

The striking thing about these new findings with the same yeast-
screening method isn't the existence of synergistic effects, but their reported magnitude. Tests using individual small doses of the estrogenic
pesticides dieldrin, toxaphene, endosulfan, and chlordane produced
little, if any, effect, but when two chemicals were combined, the response was powerfully enhanced. If further experiments confirm these
findings, this study will undoubtedly have profound implications for the
regulation of chemicals, which are now reviewed individually.

As new studies accumulate—on synergy, plant estrogens, sperm
counts, urogenital birth defects, immune system effects, and neurological hazards—they bear witness to the fact that endocrine disruption is
not a fleeting fear. This issue is here to stay, and *it is* important to the
human prospect.

In *Our Stolen Future* we raised some daunting questions, knowing
at the outset that it is not yet possible to answer all of them. We are
pleased that the book and the controversy surrounding it have served to
invigorate research. We expect that the new science emerging will, if
anything, broaden the range of plausible impacts and deepen the understanding of when, why, and how some synthetic contaminants disrupt
fetal development. It will likely reveal that some concerns are less vital
than others. It will pursue outstanding questions: How widespread is
the exposure? When and why do low doses have unexpectedly strong
impacts? What factors favor synergistic interactions? Are other unexpected sources of exposure likely? Can traditional risk assessment
assimilate hormone disruption and, if not, what can replace it?

These and many other questions remain without definitive answers, at least for now. In the meantime, even though the case has not
been proven to the satisfaction of all, we urge caution. The possible
consequences of widespread hormone disruption are immense and
irreversible.

APPENDIX:
THE WINGSPREAD
CONSENSUS STATEMENT

Authors' Note: In July 1991, a group of scientists, including Theo Colborn and Pete Myers, came together for the first time to discuss their concerns about the prevalence and effect of endocrine disrupting chemicals in the environment. That scientists from so many different disciplines were brought together in the first place is remarkable, but in the hope that their meeting might have some lasting effect they reached consensus on the following statement. We have included it here not only because it forms a succinct overview of the problem we face but also as a starting point for scientists, policy makers, and concerned individuals about the direction that research and public policy might take on this important issue. The scientists who signed this consensus are listed at the end of the statement, which follows. By including this list we do not imply that those listed (other than the present authors) necessarily endorse the arguments or conclusions presented elsewhere in this book.

Chemically-Induced Alterations in Sexual Development: The Wildlife/Human Connection

THE PROBLEM

Many compounds introduced into the environment by human activity are capable of disrupting the endocrine system of animals, including fish, wildlife, and humans. The consequences of such disruption can be profound because of the crucial role hormones play in controlling development. Because of the increasing and pervasive contamination of the environment by compounds capable of such activity, a multidisciplinary group of experts gathered in retreat at Wingspread, Racine, Wisconsin, 26–28 July 1991 to assess what is known about the issue. Participants included experts in the fields of anthropology, ecology, comparative endocrinology, histopathology, immunology, mammalogy, medicine, law, psychiatry, psychoneuroendocrinology, reproductive physiology, toxicology, wildlife management, tumor biology, and zoology.

The purposes of the meeting were:

1. to integrate and evaluate findings from the diverse research disciplines concerning the magnitude of the problem of endocrine disruptors in the environment;
2. to identify the conclusions that can be drawn with confidence from existing data; and
3. to establish a research agenda that would clarify uncertainties remaining in the field.

CONSENSUS STATEMENT

The following consensus was reached by participants at the workshop.

1. *We are certain of the following:*

   ■ A large number of man-made chemicals that have been released into the environment, as well as a few natural ones, have the potential to disrupt the endocrine system of animals, including humans. Among these are the persistent, bioaccumulative, organohalogen compounds that include some pesticides (fungicides, herbicides, and insecticides) and industrial chemicals, other synthetic products, and some metals.*

■ Many wildlife populations are already affected by these compounds. The impacts include thyroid dysfunction in birds and fish; decreased fertility in birds, fish, shellfish, and mammals; decreased hatching success in birds, fish, and turtles; gross birth deformities in birds, fish, and turtles; metabolic abnormalities in birds, fish, and mammals; behavioral abnormalities in birds; demasculinization and feminization of male fish, birds, and mammals; defeminization and masculinization of female fish and birds; and compromised immune systems in birds and mammals.

■ The patterns of effects vary among species and among compounds. Four general points can nonetheless be made: (1) the chemicals of concern may have entirely different effects on the embryo, fetus, or perinatal organism than on the adult; (2) the effects are most often manifested in offspring, not in the exposed parent; (3) the timing of exposure in the developing organism is crucial in determining its character and future potential; and (4) although critical exposure occurs during embryonic development, obvious manifestations may not occur until maturity.

■ Laboratory studies corroborate the abnormal sexual development observed in the field and provide biological mechanisms to explain the observations in wildlife.

■ Humans have been affected by compounds of this nature, too. The effect of DES (diethylstilbestrol), a synthetic therapeutic agent, like many of the compounds mentioned above, are es-

---

*Chemicals known to disrupt the endocrine system include: DDT and its degradation products, DEHP (di(2-ethylhexyl)phthalate), dicofol, HCB (hexachlorobenzene), kelthane, kepone, lindane and other hexachlorocyclohexane congeners, methoxychlor, octachlorostyrene, synthetic pyrethroids, triazine herbicides, EBDC fungicides, certain PCB congeners, 2,3,7,8-TCDD and other dioxins, 2,3,7,8-TCDF and other furans, cadmium, lead, mercury, tributyltin and other organo-tin compounds, alkyl phenols (non-biodegradable detergents and anti-oxidants present in modified polystyrene and PVCs), styrene dimers and trimers, soy products, and laboratory animal and pet food products.

trogenic. Daughters born to mothers who took DES now suffer increased rates of vaginal clear cell adenocarcinoma, various genital tract abnormalities, abnormal pregnancies, and some changes in immune responses. Both sons and daughters exposed *in utero* experience congenital anomalies of their reproductive system and reduced fertility. The effects seen in *in utero* DES-exposed humans parallel those found in contaminated wildlife and laboratory animals, suggesting that humans may be at risk to the same environmental hazards as wildlife.

2. *We estimate with confidence that:*

- Some of the developmental impairments reported in humans today are seen in adult offspring of parents exposed to synthetic hormone disruptors (agonists and antagonists) released in the environment. The concentrations of a number of synthetic sex hormone agonists and antagonists measured in the U.S. human population today are well within the range and dosages at which effects are seen in wildlife populations. In fact, experimental results are being seen at the low end of current environmental concentrations.

- Unless the environmental load of synthetic hormone disruptors is abated and controlled, large scale dysfunction at the population level is possible. The scope and potential hazard to wildlife and humans are great because of the probability of repeated and/or constant exposure to numerous synthetic chemicals that are known to be endocrine disruptors.

- As attention is focused on this problem, more parallels in wildlife, laboratory, and human research will be revealed.

3. *Current models predict that:*

- The mechanisms by which these compounds have their impact vary, but they share the general properties of (1) mimicking the effects of natural hormones by recognizing their binding sites; (2) antagonizing the effect of these hormones by blocking their interaction with their physiological binding sites; (3) reacting directly and indirectly with the hormone in question; (4) by altering the natural pattern of synthesis of hormones; or (5) altering hormone receptor levels.

- Both exogenous (external source) and endogenous (internal source) androgens (male hormones) and estrogens (female hormones) can alter the development of brain function.
- Any perturbation of the endocrine system of a developing organism may alter the development of that organism: typically these effects are irreversible. For example, many sex-related characteristics are determined hormonally during a window of time in the early stages of development and can be influenced by small changes in hormone balance. Evidence suggests that sex-related characteristics, once imprinted, may be irreversible.
- Reproductive effects reported in wildlife should be of concern to humans dependent upon the same resources, e.g., contaminated fish. Food fish is a major pathway of exposure for birds. The avian (bird) model for organochlorine endocrine disruption is the best described to date. It also provides support for the wildlife/human connection because of similarities in the development of the avian and mammalian endocrine systems.

4. *There are many uncertainties in our predictions because:*

- The nature and extent of the effects of exposure on humans are not well established. Information is limited concerning the disposition of these contaminants within humans, especially data on concentrations of contaminants in embryos. This is compounded by the lack of measurable endpoints (biologic markers of exposure and effect) and the lack of multi-generational exposure studies that simulate ambient concentrations.
- While there are adequate quantitative data concerning reduction in reproductive success in wildlife, data are less robust concerning changes in behavior. The evidence, however, is sufficient to call for immediate efforts to fill these knowledge gaps.
- The potencies of many synthetic estrogenic compounds relative to natural estrogens have not been established. This is important because contemporary blood concentrations of some of the compounds of concern exceed those of internally produced estrogens.

5. *Our judgment is that:*

- Testing of products for regulatory purposes should be broadened to include hormonal activity *in vivo*. There is no substitute for animal studies for this aspect of testing.
- Screening assays for androgenicity and estrogenicity are available for those compounds that have direct hormonal effects. Regulations should require screening all new products and by-products for hormonal activity. If the material tests positive, further testing for functional teratogenicity (loss of function rather than obvious gross birth defects) using multigenerational studies should be required. This should apply to all persistent, bioaccumulative products released in the past as well.
- It is urgent to move reproductive effects and functional teratogenicity to the forefront when evaluating health risks. The cancer paradigm is insufficient because chemicals can cause severe health effects other than cancer.
- A more comprehensive inventory of these compounds is needed as they move through commerce and are eventually released to the environment. This information must be made more accessible. Information such as this affords the opportunity to reduce exposure through containment and manipulation of food chains. Rather than separately regulating contaminants in water, air, and land, regulatory agencies should focus on the ecosystem as a whole.
- Banning the production and use of persistent chemicals has not solved the exposure problem. New approaches are needed to reduce exposure to synthetic chemicals already in the environment and prevent the release of new products with similar characteristics.
- Impacts on wildlife and laboratory animals as a result of exposure to these contaminants are of such a profound and insidious nature that a major research initiative on humans must be undertaken.
- The scientific and public health communities' general lack of awareness concerning the presence of hormonally active environmental chemicals, functional teratogenicity, and the con-

cept of transgenerational exposure must be addressed. Because functional deficits are not visible at birth and may not be fully manifested until adulthood, they are often missed by physicians, parents, and the regulatory community, and the causal agent is never identified.

6. *To improve our predictive capability:*

- More basic research in the field of developmental biology of hormonally responsive organs is needed. For example, the amount of specific endogenous hormones required to evoke a normal response must be established. Specific biologic markers of normal development per species, organ, and stage of development are needed. With this information, levels that elicit pathological changes can be established.
- Integrated cooperative research is needed to develop both wildlife and laboratory models for extrapolating risks to humans.
- The selection of a sentinel species at each trophic level in an ecosystem is needed for observing functional deficits, while at the same time describing the dynamics of a compound moving through the system.
- Measurable endpoints (biologic markers) as a result of exposure to exogenous endocrine disruptors are needed that include a range of effects at the molecular, cellular, organismal, and population levels. Molecular and cellular markers are important for the early monitoring of dysfunction. Normal levels and patterns of isoenzymes and hormones should be established.
- In mammals, exposure assessments are needed based on body burdens of a chemical that describe the concentration of a chemical in an egg (ovum) which can be extrapolated to a dose of the chemical to the embryo, fetus, newborn, and adult. Hazard evaluations are needed that repeat in the laboratory what is being seen in the field. Subsequently, a gradient of doses for particular responses must be determined in the laboratory and then compared with exposure levels in wildlife populations.
- More descriptive field research is needed to explain the annual

influx to areas of known pollution of migratory species that appear to maintain stable populations in spite of the relative vulnerability of their offspring.

- A reevaluation of the *in utero* DES-exposed population is required for a number of reasons. First, because the unregulated, large-volume releases of synthetic chemicals coincide with the use of DES, the results of the original DES studies may have been confounded by widespread exposure to other synthetic endocrine disruptors. Second, exposure to a hormone during fetal life may elevate responsiveness to the hormone during later life. As a result, the first wave of individuals exposed to DES *in utero* is just reaching the age where various cancers (vaginal, endometrial, breast, and prostatic) may start appearing if the individuals are at a greater risk because of perinatal exposure to estrogen-like compounds. A threshold for DES adverse effects is needed. Even the lowest recorded dose has given rise to vaginal adenocarcinoma. DES exposure of fetal humans may provide the most-severe-effect model in the investigation of the less potent effects from environmental estrogens. Thus, the biological endpoints determined in *in utero* DES-exposed offspring will lead the investigation in humans following possible ambient exposures.

- The effects of endocrine disruptors on longer-lived humans may not be as easily discerned as in shorter-lived laboratory or wildlife species. Therefore, early detection methods are needed to determine if human reproductive capability is declining. This is important from an individual level, as well as at the population level, because infertility is a subject of great concern and has psychological and economic impacts. Methods are now available to determine fertility rates in humans. New methods should involve more use of liver-enzyme-system activity screening, sperm counts, analyses of developmental abnormalities, and examination of histopathological lesions. These should be accompanied by more and better biomarkers of social and behavioral development, the use of multigenerational histories of individuals and their progeny, and congener-specific chemical analyses of reproductive tissues and products, including breast milk.

Work Session participants included:

Dr. Howard A. Bern
Prof. of Integrative Biology
(Emeritus) and Research
Endocrinologist
Dept. of Integrative Biology and
Cancer Research Lab
University of California
Berkeley, CA

Dr. Phyllis Blair
Prof. of Immunology
Dept. of Molecular and Cell Biology
University of California
Berkeley, CA

Sophie Brasseur
Marine Biologist
Dept. of Estuarine Ecology,
Research
Institute for Nature Management
Texel, The Netherlands

Dr. Theo Colborn
Senior Fellow
World Wildlife Fund, Inc., and
W. Alton Jones Foundation, Inc.
Washington, DC

Dr. Gerald R. Cunha
Developmental Biologist
Dept. of Anatomy
University of California
San Francisco, CA

Dr. William Davis*
Research Ecologist
U.S. EPA
Environmental Research Lab
Sabine Island, FL

Dr. Klaus D. Döhler
Director Research
Development and Production
Pharma Bissendorf Peptide GmbH
Hannover, Germany

Mr. Glen Fox
Contaminants Evaluator
National Wildlife Research Center
Environment Canada
Quebec, Canada

Dr. Michael Fry
Research Faculty
Dept. of Avian Science
University of California
Davis, CA

Dr. Earl Gray*
Section Chief
Developmental and Reproductive
Toxicology Section
Reproductive Toxicology Branch
Developmental Biology Division
Health Effects Research Laboratory
U.S. EPA
Research Triangle Park, NC

Dr. Richard Green
Prof. of Psychiatry in Residence
Dept. of Psychiatry/NPI
School of Medicine
University of California
Los Angeles, CA

Dr. Melissa Hines
Asst. Prof. in Residence
Dept. of Psychiatry/NPI
School of Medicine
University of California
Los Angeles, CA

*Although the research described in this article has been supported by the USEPA,
it does not necessarily reflect the views of the Agency and no official endorsement
should be inferred. Mention of trade names or commercial products does not constitute endorsement or recommendation for use.

Mr. Timothy J. Kubiak
Environmental Contaminants
Specialist
Dept. of Interior
U.S. Fish and Wildlife Service
East Lansing, MI

Dr. John McLachlan
Director, Div. of Intramural
Research
Chief, Laboratory of Reproductive
and Developmental Toxicology
National Institute of
Environmental
Health Sciences
National Institutes of Health
Research Triangle Park, NC

Dr. J. P. Myers
Director
W. Alton Jones Foundation, Inc.
Charlottesville, VA

Dr. Richard E. Peterson
Prof. of Toxicology and
Pharmacology
School of Pharmacy
University of Wisconsin
Madison, WI

Dr. P. J. H. Reijnders
Head, Section of Marine
Mammalogy
Dept. of Estuarine Ecology
Research Institute for Nature
Management
Texel, The Netherlands

Dr. Ana Soto
Associate Prof.
Dept. of Anatomy and Cellular
Biology
Tufts University School of
Medicine
Boston, MA

Dr. Glen Van Der Kraak
Asst. Prof.
College of Biological Sciences
Dept. of Zoology
University of Guelph
Ontario, Canada

Dr. Frederick vom Saal
Prof.
College of Arts and Sciences
Division of Biological Sciences
University of Missouri
Columbia, MO

Dr. Pat Whitten
Asst. Prof.
Dept. of Anthropology
Emory University
Atlanta, GA

NOTES

CHAPTER 1· OMENS

The paragraph beginning "In years of watching . . ." refers to C. Broley, "The Plight of the American Bald Eagle," *Audubon Magazine* 60:162–63, 171 (1958), and his wife, M. Broley, in *Eagleman*, Pellegrini and Cudahy, 1952.

"Although otters . . ." is based on writings by Chris Mason, including C. Mason, T. Ford, and N. Last, "Organochlorine Residues in British Otters," *Bulletin of Environmental Contamination and Toxicology* 36:656–61 (1986); and C. Mason and S. Macdonald, *Otters: Ecology and Conservation*, Cambridge University Press, 1986.

"In the post–World War II . . ." refers to a series of papers by Richard Aulerich and Robert Ringer, that includes R. Aulerich, R. Ringer, and S. Iwamoto, "Reproductive Failure and Mortality in Mink Fed on Great Lakes Fish," *Journal of Reproduction and Fertility Supplement* 19:365–76 (1973).

"Curiously, other mink ranchers . . ." comes from D. Dutton, *Worse Than the Disease: Pitfalls of Medical Progress*, Cambridge University Press, 1988.

"The sight of . . ." is taken from personal conversations with Michael Gilbertson and a paper by M. Gilbertson, T. Kubiak, J. Ludwig, and G. Fox, "Great Lakes Embryo Mortality, Edema, and Deformities Syndrome (GLEMEDS) in Colonial Fish-Eating Birds: Similarity to Chick-Edema Disease," *Journal of Toxicology and Environmental Health* 33(4):455–520 (1991).

"Even the trained eye . . ." leans on work from a number of researchers: G. Hunt and M. Hunt, "Female-Female Pairing in Western Gulls (*Larus occidentalis*) in Southern California," *Science* 196:1466–67 (1977); M. Conover and G. Hunt, "Female-Female Pairing and Sex Ratios in Gulls: An Historical Perspective," *Wilson Bulletin*, 96(4):619–25 (1984); D. Fry and M. Toone, "DDT-Induced Feminization of Gull Embryos," *Science* 213:922–24 (1981); G. Fox, S. Teeple, A. Gilman, F. Anderka, and G. Hogan, "Are Lake Ontario Herring Gulls Good Parents?" in *Proceedings of the Fish-Eating Birds of the Great Lakes and Environmental Contaminants Symposium, December 2–3, 1976*, Co-sponsored by the Toxic Chemical Division and Ontario Region, Canadian Wildlife Service, pp. 76–90 (1976); and personal communication (1994) with Ian Nisbet, North Falmouth, and Jeremy Hatch, University of Massachusetts, Boston.

The section about Florida alligators stems from: A. Woodward, H. Percival, M. Jennings, and C. Moore, "Low Clutch Viability of American Alligators on Lake Apopka," *Florida Science* 56:52–63 (1993); and L. Guillette, T. Gross, D. Gross, A. Rooney, and H. Percival, "Gonadal Steroidogenesis *In Vitro* from Juvenile Alligators Obtained from Contaminated or Control Lakes," *Environmental Health Perspectives* 103(4):31–36 (1995). This section was also supported by personal communications with Louis Guillette, University of Florida, Gainesville, 1994–95.

"The first signs of the epidemic . . ." relates to a series of events described in R. Dietz, M.-P. Heide-Jørgensen, and T. Härkönen, "Mass Deaths of Harbor Seals (*Phoca vitulina*) in Europe," *Ambio* 18(5):258–264 (1989).

The section about dolphins beginning "Although fishermen and yachtsmen . . ." discusses the research of Alex Aguilar and a number of collaborators from around the world. See A. Aguilar and J. Raga, "The Striped Dolphin Epizootic in the Mediterranean Sea," *Ambio* 22(8):524–28 (1993); K. Kannan, S. Tanabe, A. Borrell, A. Aguilar, S. Focardi, and R. Tatsukawa, "Isomer-Specific Analysis and Toxic Evaluation of Polychlorinated Biphenyls in Striped Dolphins Affected by an Epizootic in the Western Mediterranean Sea," *Archives of Environmental Contamination and Toxicology* 25:227–33 (1993); J. Forcada, A. Aguilar, P. Hammond, X. Pastor, and R. Aguilar, "Distribution and Numbers of Striped Dolphins in the Western Mediterranean Sea After the 1990 Epizootic Outbreak," *Marine Mammal Science* 10(2):137–50 (1994); and A. Aguilar and A. Borrell, "Abnormally High Polychlorinated Biphenyl Levels in Striped Dolphins (*Stenella coeruleoalba*) Affected by the 1990–1992 Mediterranean Epizootic," *The Science of the Total Environment* 154:237–47 (1994).

"Even a high school . . ." refers to E. Carlsen, A. Giwercman, N. Keiding, and N. Skakkebaek, "Evidence for Decreasing Quality of Semen During Past 50 Years," *British Medical Journal* 305:609–13 (1992).

Chapter 2: HAND-ME-DOWN POISONS

The paper about herring gulls is R. Moccia, G. Fox, A. Britton, *The Journal of Wildlife Diseases* 22(1):60–70 (1986).

For further information concerning the paragraph beginning "Flick. Another document . . . ," see T. Colborn, A. Davidson, S. Green, R. Hodge, C. Jackson, and R. Liroff, *Great Lakes, Great Legacy?*, The Conservation Foundation and the Institute for Research on Public Policy, 1990; and G. Fox, "What Have Biomarkers Told Us About the Effects of Contaminants on the Health of Fish-Eating Birds in the Great Lakes? The Theory and a Literature Review," *Journal of Great Lakes Research* 19(4):722–36 (1993).

For more information about opportunistic species, loss of diversity, and overloading of a system by rapidly reproducing organisms alluded to starting with "The improvements since then . . . ," see D. Rapport, H. Regier, and T. Hutchinson, "Ecosystem Behavior Under Stress," *The American Naturalist* 125(5):617–38 (1985); H. Regier and G. Baskerville, "Sustainable Redevelopment of Degraded Ecosystems," in *Sustainable Development of the Biosphere*, W. Clark and R. Munn, eds., Cambridge University Press, 1986. The explosion of double-crested cormorants, not only in the Great Lakes, but across the United States after DDT use was curtailed, is an example of a species that took off like a weed, filling niches that were left vacant by more sensitive species. "She had already . . ." stems from general opinion polls in Canada and from the joint report by the National Research Council of the United States and The Royal Society of Canada, "The Great Lakes Water Quality Agreement: An Evolving Instrument for Ecosystem Management," 1985.

"John Harshbarger . . ." refers to a series of papers that include: P. Baumann and J. Harshbarger, "Frequencies of Liver Neoplasia in a Feral Fish Population and Associated Carcinogens," *Marine Environmental Research* 17:324–27 (1985); J. Black, "Epidermal Hyperplasia and Neoplasia in Brown Bullheads (*Ictalurus nebulosus*) in Response to Repeated Applications of a PAH Containing Extract of Polluted River Sediment," in *Polynuclear Aromatic Hydrocarbons: Seventh International Symposium on Formation, Metabolism and Measurement*, M. Cooke and A. Dennis, eds., Battelle, 1982, pp. 99–111; A. Maccubbin, P. Black, L. Trzeciak, and J. Black, "Evidence for Polynuclear Aromatic Hydrocarbons in the Diet of Bottom-Feeding Fish," *Bulletin of Environmental Contamination and Toxicology* 34:876–82 (1985).

"Those trying to discover . . ." mentions a keynote address at the Toronto meeting by B.-E. Bengtsson. For further reading see B.-E. Bengtsson, A. Bergman, I. Brandt, C. Hill, N. Johansson, A. Södergren, and J. Thulin,

"Reproductive Disturbances in Baltic Fish: Research Programme for the Period 1994/95 to 1997/98" in a report for the Swedish Environmental Protection Agency, 1994; and L. Norrgren, "Report from the Uppsala Workshop on Reproduction Disturbances in Fish," Report #4346, Swedish Environmental Protection Agency, Research and Development Department, 1994.

The paragraphs that follow "In herring gull . . ." explain in more detail the episodes cited in Chapter 1 under the Channel Islands, Southern California, episode; see those citations.

"Earlier experiments by other . . ." discusses data fully covered in D. Fry, C. Toone, S. Speich, and R. Peard, "Sex Ratio Skew and Breeding Patterns of Gulls: Demographic and Toxicological Considerations," *Studies in Avian Biology* 10:26–43 (1987); and T. Kubiak, H. Harris, L. Smith, T. Schwartz, D. Stalling, J. Trick, L. Sileo, D. Docherty, and T. Erdman, "Microcontaminants and Reproductive Impairment of the Forster's Tern on Green Bay, Lake Michigan—1983," *Archives of Environmental Contamination and Toxicology* 18:706–27 (1989).

"Fox and others . . ." refers to G. Fox, A. Gilman, D. Peakall, F. Anderka, "Behavioral Abnormalities of Nesting Lake Ontario Herring Gulls," *Journal of Wildlife Management* 42(3):477–83 (1978).

The comment about "gay gulls" (in the paragraph that starts with "As Colborn tackled . . .") refers to J. Diamond, "Goslings of Gay Geese," *Nature* 340:101 (1989). In this article Diamond points out that female-female pairing had never been reported in the literature before the 1950s.

The endocrinology textbook referred to is G. Hedge, H. Colby, and R. Goodman, *Clinical Endocrine Physiology*, W. B. Saunders, 1987.

"It dawned on Colborn . . ." refers to a series of reports, some of which are: G. Fein, J. Jacobson, S. Jacobson, P. Schwartz, and J. Dowler, "Prenatal Exposure to Polychlorinated Biphenyls: Effects on Birth Size and Gestational Age," *Journal of Pediatrics* 105(2):315–20 (1984); S. Jacobson, G. Fein, J. Jacobson, P. Schwartz, and J. Dowler, "The Effect of Intrauterine PCB Exposure on Visual Recognition Memory," *Child Development* 56:853–60 (1985); J. Jacobson, S. Jacobson, and H. Humphrey, "Effects of In Utero Exposure to Polychlorinated Biphenyls and Related Contamination on Cognitive Functioning in Young Children," *Journal of Pediatrics* 116:38–45 (1990); and J. Jacobson, S. Jacobson, and H. Humphrey, "Effects of Exposure to PCBs and Related Compounds on Growth and Activity in Children," *Neurotoxicology and Teratology* 12:319–26 (1990).

For a comprehensive overview of chemicals found in breast milk, see A. Jensen and S. Slorach, *Chemical Contaminants in Human Milk*, CRC Press, 1991. Also see K. Thomas and T. Colborn, "Organochlorine Endocrine

Disruptors in Human Tissue," in *Chemically Induced Alterations in Sexual and Functional Development: The Wildlife–Human Connection*, T. Colborn and C. Clement, eds., Princeton Scientific Publishing, 1992, pp. 365–94.

"Of course! . . ." discusses the problem of biomagnification; see R. Norstrom, D. Hallett, and R. Sonstegard, "Coho Salmon (*Oncorhynchus kisutch*) and Herring Gulls (*Larus arentatus*) as Indicators of Organochlorine Contamination in Lake Ontario," *Journal of the Fisheries Research Board of Canada* 35(11):1401–1409 (1978). These authors reported that PCBs biomagnified 25 million times from Lake Ontario water to the herring gull.

The foundation for this chapter was Colborn's report under contract to Darrell Piekarz, Environment Canada, Conservation and Protection, Environmental Interpretation Division, Ottawa, Canada, "The Great Lakes Toxics Working Paper," Contract Number: KE144-7-6336, April 19, 1988.

CHAPTER 3: CHEMICAL MESSENGERS

For the reader who wants to delve deeper into the major theme of this chapter, we recommend: F. vom Saal, "The Intrauterine Position Phenomenon. Effects on Physiology, Aggressive Behavior, and Population Dynamics in House Mice," in *Biological Perspectives on Aggression*, K. Flannelly, R. Blanchard, and D. Blanchard, eds. (no. 169 in a series entitled *Progress in Clinical Biology Research*); Liss, 1984, pp. 135–79; F. vom Saal, "Sexual Differentiation in Litter Bearing Mammals: Influence of Sex of Adjacent Fetuses in Utero," *Journal of Animal Science* 67:1824–40 (1989); and F. vom Saal, M. Montano, and M. Wang, "Sexual Differentiation in Mammals," in *Chemically Induced Alterations in Sexual and Functional Development: The Wildlife–Human Connection*, T. Colborn and C. Clement, eds., Princeton Scientific Publishing, 1992, pp. 17–83.

"The sisters also . . ." refers to F. vom Saal and F. Bronson, "Sexual Characteristics of Adult Female Mice Are Correlated with Their Blood Testosterone Levels During Prenatal Development," *Science* 208:597–99 (1980).

"Not so fast . . ." is supported by the vom Saal paper on intrauterine position and J. Vandenbergh, "Regulation of Puberty and Its Consequences on Population Dynamics of Mice," *American Zoologist* 27:891–98 (1987).

"Even more amazing . . ." refers to a report in the Science Section of the *New York Times*, Tuesday, March 31, 1992, "Prenatal Womb Position and Supermasculinity," in which Bennett Galef and coworkers, working with Mongolian gerbils, are quoted as stating that "almost everything we've looked at behaviorally is affected by intrauterine position." See also

M. Clark, P. Karpiuk, and B. Galef, "Hormonally Mediated Inheritance of Acquired Characteristics in Mongolian Gerbils," *Nature* 364:712 (1993).

"Interestingly, some studies . . ." refers to F. vom Saal, D. Quadagno, M. Even, L. Keisler, D. Keisler, and S. Khan, "Paradoxical Effects of Maternal Stress on Fetal Steroids and Postnatal Reproductive Traits in Female Mice from Different Intrauterine Positions," *Biology of Reproduction* 43: 751–61 (1990).

"Whatever the source . . ." cites D. McFadden, "A Masculinizing Effect on the Auditory Systems of Human Females Having Male Co-Twins," *Proceedings of the National Academy of Science* 90:11900–11904 (1993).

The calculations in the paragraph starting "The striking . . ." are based on a standard medicine dropper delivering 20 drops per milliliter or cubic centimeter.

For further reading related to the paragraph starting with "In girls, the change . . . ," see F. vom Saal, C. Finch, and J. Nelson, "Natural History and Mechanisms of Reproductive Aging in Humans, Laboratory Rodents, and Other Selected Vertebrates" in *The Physiology of Reproduction*, 2nd ed., E. Knobil and J. Neill, eds., Plenum, 1994, pp. 1213–1314.

"An individual who gets . . ." mentions Charles Phoenix, who also worked with Robert Goy. See C. Phoenix, R. Goy, A. Gerall, and W. Young, "Organizing Action of Prenatally Administered Testosterone Propionate on the Tissues Mediating Mating Behavior in the Female Guinea Pig," *Endocrinology* 65:369–82 (1959).

To read more about sexual differentiation, see *Behavioral Endocrinology*, J. Becker, S. Marc, and D. Crews, eds., MIT Press, 1993; and S. LeVay, *The Sexual Brain*, MIT Press, 1993.

CHAPTER 4: HORMONE HAVOC

"From the very beginning . . ." describes the work of R. Greene, M. Burrill, and A. Ivy, "Experimental Intersexuality: The Paradoxical Effects of Estrogens on the Sexual Development of the Female Rat," *Anatomical Record* 74(4):429–38 (1939); and R. Greene, M. Burrill, and A. Ivy, "Experimental Intersexuality: Modification of Sexual Development of the White Rat with a Synthetic Estrogen," in *Proceedings of the Society for Experimental Biology and Medicine* 41:169–70 (1939).

For further reading about the passage starting "This cautionary evidence . . . ," see D. Dutton, *Worse Than the Disease: Pitfalls of Medical Progress*, Cambridge University Press, 1988.

"The Northwestern University rat studies . . ." reflects a conversation with John McLachlan on May 5, 1994, who at that time was chief at the Laboratory of Reproductive and Developmental Toxicology, National Institute of Environmental Health Sciences.

"Before doctors . . ." refers to Insight Team of the *Sunday Times* of London, *Suffer the Children: The Story of Thalidomide*, Viking, 1979.

"As would later . . ." also refers to the Insight Team's book.

"For ordinary people . . ." refers to "The Full Story of the Drug Thalidomide," *Life* magazine, August 10, 1962.

"Would doctors have . . ." was reenforced by the discussion with John McLachlan on May 5, 1994, mentioned earlier.

"The most painful aspect . . ." refers to W. Dieckmann, M. Davis, L. Rynkiewicz, and R. Pottinger, "Does the Administration of Diethylstilbestrol During Pregnancy Have Therapeutic Value?" *American Journal of Obstetrics and Gynecology* 66(5):1062 (1953); and Y. Brackbill and H. Berendes, "Dangers of Diethylstilboestrol: Review of a 1953 Paper," a letter in *Lancet* 2:520 (1978).

The closing of the paragraph that begins "When the cluster . . ." refers to the book by D. Dutton.

"Ulfelder just couldn't . . ." describes a scene from R. Meyers, *D. E. S.: The Bitter Pill*, Seaview/Putnam, 1983, pp. 93–94.

The paragraph that follows refers to A. Herbst, H. Ulfelder, and D. Poskanzer, "Adenocarcinoma of the Vagina: Association of Maternal Stilbestrol Therapy with Tumor Appearance in Young Women," *New England Journal of Medicine* 284:878–81 (1971). Herbst, who played an invaluable role in bringing the DES tragedy to light, continues to follow the medical histories of the DES offspring.

"Regardless of the compelling . . ." quotes T. Dunn and A. Green, "Cysts of the Epididymis, Cancer of the Cervix, Granular Cell Myoblastoma, and Other Lesions After Estrogen Injection in Newborn Mice," *Journal of the National Cancer Institute* 31:425–38 (1963); A. Herbst and H. Bern, eds., *Developmental Effects of Diethylstilbestrol (DES) in Pregnancy*, Thieme-Stratton, 1981, p. 1; and N. Takasugi and H. Bern, "Tissue Changes in Mice with Persistent Vaginal Cornification Induced by Early Postnatal Treatment with Estrogen," *Journal of the National Cancer Institute* 33:855–65 (1964).

"Before long, the group . . ." refers to R. Newbold and J. McLachlan, "Vaginal Adenosis and Adenocarcinoma in Mice Exposed Prenatally or Neonatally to Diethylstilbestrol," *Cancer Research* 42:2003–11 (1982).

"McLachlan and his colleagues . . ." cites J. McLachlan, R. Newbold, and B. Bullock, "Reproductive Tract Lesions in Male Mice Exposed Prenatally to Diethylstilbestrol," *Science* 190:991–92 (1975).

"No doubt because . . ." refers to W. Gill, "Effects on Human Males of *In-Utero* Exposure to Exogenous Sex Hormones," in *Toxicity of Hormones in Perinatal Life*, T. Mori and H. Nagasawa, eds., CRC Press, 1988, pp. 162–74.

"Friedman pursued his hunch . . ." cites B. Tilley, A. Barnes, E. Bergstralh, D. Labarthe, K. Noller, T. Colton, and E. Adam, "A Comparison of Maternal History Recall and Medical Records: Implications for Retrospective Studies," *American Journal of Epidemiology*, 121(2):269–81 (1985).

"Friedman decided to . . ." describes the frustrations mentioned in D. Schottenfeld, M. Warshauer, S. Sherlock, A. Zauber, M. Leder, and R. Payne, "The Epidemiology of Testicular Cancer in Young Adults," *American Journal of Epidemiology*, 112(2):232–46 (1980).

For more information about the paragraph opening "As was the case . . . ," see C. Orenberg, *D. E. S.: The Complete Story*, St. Martin's, 1981, pp. 46–47.

"In immune system . . ." discusses T cells, which comprise a small part of the total population of white blood cells (lymphocytes) in the immune system. T cells get their name because they differentiate in the thymus gland whereas the other cells in the immune system differentiate in the fetal liver, spleen, and adult bone marrow. T cells function as the first line of defense against viruses and other foreign material in the body. Among their number are the natural killer cells (NK) that target and kill cells that have been attacked by viruses or have been transformed into cancerous cells.

After reading "Although DES exposed mice . . ." the reader might want to consult I. Palmlund, R. Apfel, S. Buitendijk, A. Cabau, and J. Forsberg, "Effects of Diethylstilbestrol (DES) Medication During Pregnancy: Report From a Symposium at the 10th International Congress of ISPOG," *Journal of Psychosomatic Obstetrical Gynaecology* 14:71–89 (1993). This paragraph also mentions a possible link between DES prenatal exposure and rheumatic fever. For more information see P. Blair, "Immunologic Studies of Women Exposed *In Utero* to Diethylstilbestrol," in *Chemically Induced Alterations in Sexual and Functional Development: The Wildlife–Human Connection*, T. Colborn and C. Clement, eds., Princeton Scientific Publishing, 1992, pp. 289–93; and P. Blair, K. Noller, J. Turiel, B. Forghani, and S. Hagens, "Disease Patterns and Antibody Responses to Viral Antigens in Women Exposed *In Utero* to Diethylstilbestrol," in *Chemically Induced Alterations*, pp. 283–88. Also see D. Wingard and J. Turiel, "Long-Term

Effects of Exposure to Diethylstilbestrol," *Journal of Western Medicine* 149:551–54 (1988).

Much of what follows "The animal studies . . ." leans heavily on M. Hines, "Surrounded by Estrogens? Considerations for Neurobehavioral Development in Human Beings," in *Chemically Induced Alterations*, pp. 261–81.

It is important to point out that other hormones have been used for the same purpose as DES, such as progesterone and testosterone, although their effects in offspring have not been as carefully documented.

CHAPTER 5: FIFTY WAYS TO LOSE YOUR FERTILITY

"Twelve years after . . ." refers to H. Burlington and V. Lindeman, "Effect of DDT on Testes and Secondary Sex Characters of White Leghorn Cockerels," *Proceedings of the Society for Experimental Biology and Medicine* 74:48–51 (1950).

"The body has hundreds . . ." discusses orphan receptors studied by a team at the National Institutes of Environmental Health Sciences. See J. McLachlan, R. Newbold, C. Teng, and K. Korach, "Environmental Estrogens: Orphan Receptors and Genetic Imprinting," in *Chemically Induced Alterations in Sexual and Functional Development: The Wildlife-Human Connection*, T. Colborn and C. Clement, eds., Princeton Scientific Publishing, 1992, pp. 107–12.

"They fit together . . ." describes mechanisms of hormone binding. See K. Korach, P. Sarver, K. Chae, J. McLachlan, and J. McKinney, "Estrogen Receptor-Binding Activity of Polychlorinated Hydroxybiphenyls: Conformationally Restricted Structural Probes," *Molecular Pharmacology* 33:120–26 (1987); and J. McLachlan, "Functional Toxicology: A New Approach to Detect Biologically Active Xenobiotics," *Environmental Health Perspectives* 101(5):386–87 (1993).

For those who want to read more about the sensitivity of the fetus, we suggest H. Bern, "The Fragile Fetus" in *Chemically Induced Alterations*, pp. 9–15.

"As scientists have explored . . ." refers to A. Salhanick, C. Vito, and T. Fox, "Estrogen-Binding Proteins in the Oviduct of the Turtle, *Chrysemys picta*: Evidence for a Receptor Species," *Endocrinology*, 105(6):1388–95 (1979).

The Australian sheep incident described in the paragraph starting "The early 1940s . . ." comes from H. Bennetts, E. Underwood, and F. Shier, "A Specific Breeding Problem of Sheep on Subterranean Clover Pastures in Western

Australia," *Australian Veterinary Journal* 22:2–12 (1946); personal communication with Norman Adams, CSIRO Division of Animal Production, PO Wembley, Australia; and N. Adams, "Organizational and Activational Effects of Phytoestrogens on the Reproductive Tract of the Ewe" (in press).

"Surprisingly, plant evolution . . ." leans on K. Setchell, "Naturally Occurring Non-Steroidal Estrogens of Dietary Origin," in *Estrogens in the Environment II: Influences on Development*, J. McLachlan, ed., Elsevier, 1985; and R. Bradbury and D. White, "Estrogens and Related Substances in Plants," *Vitamins and Hormones* 12:207–33 (1954).

"For such a defensive strategy . . ." refers to C. Hughes, "Phytochemical Mimicry of Reproductive Hormones and Modulation of Herbivore Fertility by Phytoestrogens," *Environmental Health Perspectives* 78:171–75 (1988).

"Humans long ago . . ." discusses J. M. Riddle, *Contraception and Abortion from the Ancient World to the Renaissance*, Harvard University Press, 1994.

"There is no . . ." begins several pages about the work of Pat Whitten. See P. Whitten, "Chemical Revolution to Sexual Revolution: Historical Changes in Human Reproductive Development," in *Chemically Induced Alterations*, pp. 311–34; P. Whitten and F. Naftolin, "Effects of a Phytoestrogen Diet on Estrogen-Dependent Reproductive Processes in Immature Female Rats," *Steroids* 57:56–61 (1992); P. Whitten, E. Russell, and F. Naftolin, "Effects of a Normal, Human-Concentration, Phytoestrogen Diet on Rat Uterine Growth," *Steroids* 57:98–106 (1992); and P. Whitten, C. Lewis, and F. Naftolin, "A Phytoestrogen Diet Induces the Premature Anovulatory Syndrome in Lactationally Exposed Female Rats," *Biology of Reproduction* 49:1117–21 (1993).

"When male workers in a chemical plant . . ." refers to a discovery cited in P. Guzelian, "Fourteen Workers Exposed to Pesticide Kepone Are Probably Sterile, Researchers Report," *Occupational Health and Safety Letters* 6:2 (1976).

"To date, researchers . . ." refers to T. Colborn, F. vom Saal, and A. Soto, "Developmental Effects of Endocrine-Disrupting Chemicals in Wildlife and Humans," *Environmental Health Perspectives* 101(5):378–84 (1993). This paragraph also introduces three synthetic chemicals that will be addressed again and again throughout this book: PCBs, dioxins, and furans. Chemicals are grouped in these families because they share a common chemical structure, but each family member differs in the arrangement of the chlorine atoms on its common structure. These family members are called *congeners*. The most famous of the dioxins, because of its toxicity, 2,3,7,8-TCDD (tetrachlorodibenzo-para-dioxin), is what is commonly referred to as "dioxin." The other members of the dioxin family and PCBs and furans are referred to by their respective chemical structure. Furans are similar in structure to dioxins but are of lower toxicity.

"Most discussions . . ." refers to the problem caused by substituting inadequately tested new chemicals for old-time chemicals that are now known to be toxic problems. For example, see A. Murk, J. van den Berg, J. Koeman, and A. Brouwer, "The Toxicity of Tetrachlorobenzyltoluenes (Ugilec 141) and Polychlorobiphenyls (Aroclor 1254 and PCB-77) Compared in Ah-Responsive and Ah-Nonresponsive Mice," *Environmental Pollution* 72:57–67 (1991). In this study, the researchers found that a German substitute for PCBs, Ugilec 141, was bioaccumulating in fish in the Rhine River and is as toxic as the products it was designed to replace. This would not have been discovered without applying forensic science in the field: traditional monitoring would have missed this product in the fish.

In reading the paragraph starting "As the number . . . ," bear in mind that the published 1993 number of chemicals (51) with endocrine or reproductive effects has already increased, although not all of the new discoveries have been reported in the scientific literature to date (1995)

The section that opens "If some scientists . . ." leans on the research of a team of reproductive toxicologists at the U.S. Environmental Protection Agency's Health Effects Research Laboratory, North Carolina. See L. Gray, J. Ostby, and W. Kelce, "Developmental Effects of an Environmental Antiandrogen: The Fungicide Vinclozolin Alters Sex Differentiation of the Male Rat," *Toxicology and Applied Pharmacology* 129:46–52 (1994); and W. Kelce, C. Stone, S. Laws, L. Gray, J. Kemppainen, and E. Wilson, "Persistent DDT Metabolite p,p'-DDE is a Potent Androgen Receptor Antagonist," *Nature* 375:581–85 (1995).

CHAPTER 6: TO THE ENDS OF THE EARTH

Guidance for polar bear natural history and behavior came from I. Stirling, *Polar Bears*, University of Michigan Press, 1988; and Thor Larsen, "Polar Bear Denning and Cub Production in Svalbard, Norway," *Journal of Wildlife Management* 49(2):320–26 (1985).

"Based on what . . ." is based on an Associated Press report by Doug Mellgren, "Norwegian Researchers Fear PCBs Threaten Polar Bears' Fertility," January 4, 1993.

"Some Svalbard bears . . ." refers to G. Norheim, J. Skaare, and Ø. Wiig, "Some Heavy Metals, Essential Elements, and Chlorinated Hydrocarbons in Polar Bear (*Ursus maritimus*) at Svalbard," *Environmental Pollution* 77(1):51–57 (1992).

"The story of . . ." uses the term *persistent*, which is based on a definition by the U.S. and Canadian International Joint Commission in their Sixth Biennial Report on Great Lakes Water Quality, 1992: "Any toxic substance with a half-life in water of greater than eight weeks." (p. 26) They recommend that the definition apply to all media: water, air, sediment, soil, and biota.

"The person to first . . ." refers to "Report of a New Chemical Hazard," a news item in *New Scientist* 32:612 (1966).

"Our imaginary PCB molecule . . ." introduces PCB congener 153. The following are selected papers that describe in more detail the nature of PCB congener 153: L. Hansen, "Environmental Toxicology of Polychlorinated Biphenyls," *Polychlorinated Biphenyls (PCBs): Mammalian and Environmental Toxicology*, S. Safe, ed., Springer-Verlag, 1987, pp. 15–48; D. Ness, S. Schantz, J. Moshtaghian, and L. Hansen, "Effects of Perinatal Exposure to Specific PCB Congeners on Thyroid Hormone Concentrations and Thyroid Histology in the Rat," *Toxicology Letters* 68:311–23 (1993) [Thyroid hormone production in rat pups whose mothers were exposed to PCB-153 was depressed. The pups' brain and body weights were lower and their livers were larger than unexposed pups.]; and B. Bush, A. Bennett, and J. Snow, "Polychlorobiphenyl Congeners, p,p'-DDE, and Sperm Function in Humans," *Archives of Environmental Contamination and Toxicology* 15:333–41 (1986) [In this study, three PCB congeners (153, 138, and 114) were inversely associated with sperm motility in samples from men with less than 20 million sperm per milliliter.].

Much of what appears in the paragraph starting "Our imaginary PCB molecule . . ." was taken from the record of Federal Civil Action 1:92-CV-2137: *Robert K. Joiner and Karen P. Joiner v. General Electric Company, Westinghouse Electric Company, and Monsanto Company*: Deposition of retired Monsanto employee William B. Papageorge, July 22, 1993.

"These ubiquitous metal . . ." relies on conversations with Diane Herndon of the public information center at Monsanto, other Monsanto officials, and Edward Bates, Jr., an engineer and manager of transformer tests at General Electric's power transformer plant, Pittsfield, Massachusetts, from 1940 until his retirement in the mid-1980s. All were invaluable in helping us reconstruct as much as possible the details of how Aroclors were used in GE's manufacturing process in Pittsfield. After all these years, some of the precise details were impossible to establish.

"Within hours . . ." describes bioaccumulation as in S. Hooper, C. Pettigrew, and G. Sayler, "Ecological Fate, Effects and Prospects for the Elimination of Environmental Polychlorinated Biphenyls (PCBs)," *Environmental Toxicology and Chemistry* 9:655–67 (1990).

As with so many species, there is concern over the decline of the American eel mentioned in the paragraph starting "Before long. . . ." For a comprehensive account of the status of the American eel in the St. Lawrence River see M. Castonguay, P. Hodson, C. Couillard, M. Eckersley, J.-D. Dutil, and G. Verreault, "Why Is Recruitment of the American Eel, *Anguilla rostrata*, Declining in the St. Lawrence River and Gulf?" *Canadian Journal of Fisheries and Aquatic Sciences* 51:479–88 (1994).

The section starting "The eel's flesh . . ." is based on the following: H. Iwata, S. Tanabe, N. Sakai, and R. Tatsukawa, "Distribution of Persistent Organochlorines in Oceanic Air and Surface Seawater and the Role of Ocean in Their Global Transport and Fate," *Environmental Science and Technology* 27(6):1080–98 (1993); F. Wania and D. Mackay, "Global Fractionation and Cold Condensation of Low Volatility Organochlorine Compounds in Polar Regions," *Ambio* 22(1):10–18 (1993), and M. Oehme, "Further Evidence for Long-range Air Transport of Polychlorinated Aromates and Pesticides. North American and Eurasia to the Arctic," *Ambio* 20(7):293–97 (1991).

"The gray green . . ." depends upon information in D. Muir, R. Norstrom, M. Simon, "Organochlorine Contaminants in Arctic Marine Food Chains: Accumulation of Specific Polychlorinated Biphenyls and Chlordane-Related Compounds," *Environmental Science and Technology* 22(9):1071–79, (1988) and personal communication with Derek Muir, 1995.

"While prenatal exposure . . ." raises questions concerning breast milk. See A. Smith, "Infant Exposure Assessment for Breast Milk Dioxins and Furans Derived from Waste Incineration Emissions," *Risk Analysis* 7(3):347–53 (1987).

"The contamination of . . ." reflects É. Dewailly, A. Nantel, J. Weber, and F. Meyer, "High Levels of PCBs in Breast Milk of Inuit Women from Arctic Quebec," *Bulletin of Environmental Contamination and Toxicology* 43:641–46 (1989); É. Dewailly, P. Ayotte, S. Bruneau, C. LaLiberté, D. Muir, and R. Norstrom, "Human Exposure to Polychlorinated Biphenyls Through the Aquatic Food Chain in the Arctic," Dioxin '93: 13th International Symposium on Chlorinated Dioxins and Related Compound, Vienna, September 1993, 14:173–75 (1993); and D. Kinloch, H. Kuhnlein, and D. Muir, "Inuit Foods and Diet: A Preliminary Assessment of Benefits and Risks," *The Science of the Total Environment* 122:247–78 (1992).

Probably the most disturbing information concerning exposure to persistent chemicals in the Arctic is revealed in D. Gregor, A. Peters, C. Teixeira,

N. Jones, and C. Spencer, "The Historical Residue Trend of PCBs in the Agassiz Ice Cap, Ellesmere Island, Canada," *The Science of the Total Environment* 160/161:117–26 (1995). The authors found there was no change in average PCB deposition between 1963 and 1993 on Ellesmere Island situated west of Greenland and 500 miles from the North Pole. PCB-153 was among the PCB congeners measured in this study.

CHAPTER 7: A SINGLE HIT

For more information about the work mentioned in the paragraph starting "A few months later . . . ," see R. Peterson, R. Moore, T. Mably, D. Bjerke, and R. Goy, "Male Reproductive System Ontogeny: Effects of Perinatal Exposure to 2,3,7,8-Tetrachlorodibenzo-p-dioxin" in *Chemically Induced Alterations in Sexual and Functional Development: The Wildlife-Human Connection*, T. Colborn and C. Clement, eds., Princeton Scientific Publishing, 1992, pp. 175–93.

The lay reader who wishes to know more about dioxin (known as 2,3,7,8-TCDD for short) should consult "Putting the Lid on Dioxins: Protecting Human Health and the Environment," a joint report by Physicians for Social Responsibility and the Environmental Defense Fund, 1994. For those with a technical background, we recommend examining the six-volume External Review Draft Dioxin Reassessment, released by the U.S. Environmental Protection Agency's Office of Research and Development in June 1994, EPA/600/BP–92/001; S. Safe, "Comparative Toxicology and Mechanism of Action of Polychlorinated Dibenzo-p-dioxins and Dibenzofurans," *Annual Review of Pharmacology and Toxicology* 26:371–99 (1986); and S. Safe, "Polychlorinated Biphenyls (PCBs), Dibenzo-p-dioxins (PCDDs), Dibenzofurans (PCDFs), and Related Compounds: Environmental and Mechanistic Considerations Which Support the Development of Toxic Equivalency Factors (TEFs)," *Critical Reviews in Toxicology* 21(1):51–88 (1990).

Figures in this chapter for production and releases of dioxin-contaminated material were taken from "Veterans and Agent Orange: Health Effects of Herbicides Used in Vietnam," Institute of Medicine, National Academy of Sciences, 1993; M. Gough, *Dioxin, Agent Orange: The Facts*, Plenum, 1986; and "The Health Risks of Dioxin" hearing before the Human Resources and Intergovernmental Relations Subcommittee of the Committee on Government Operations, House of Representatives, June 10, 1992.

For the reader interested in following the cancer studies from Severo, we recommend P. Bertazzi, A. Pesatori, D. Consonni, A. Tironi, M. Landi, and

C. Zocchetti, "Cancer Incidence in a Population Accidentally Exposed to 2,3,7,8-Tetrachlorodibenzo-para-dioxin," *Epidemiology* 4(5):398–406 (1993). We further recommend A. Pesatori, D. Consonni, A. Tironi, C. Zocchetti, and P. Bertazzi, "Cancer in a Young Population in a Dioxin-Contaminated Area," *International Journal of Epidemiology* 22(6):1010–13 (1993); and P. Bertazzi, C. Zocchetti, A. Pesatori, S. Guercilena, M. Sanarico, and L. Radice, "Ten-Year Mortality Study of the Population Involved in the Seveso Incident in 1976," *American Journal of Epidemiology* 129(6):1187–1200 (1989). The Pesatori et al., 1993, study revealed an increase in thyroid cancer, which, as the authors state, is consistent with "experimental findings and previous observations in humans." (p. 1010)

"In this case . . ." refers to G. Smoger, P. Kahn, G. Rodgers, S. Suffin, and P. McConnachie, "*In Utero* and Postnatal Exposure to 2,3,7,8-TCDD in Times Beach, Missouri: 1. Immunological Effects: Lymphocyte Phenotype Frequencies," Dioxin '93: 13th International Symposium on Chlorinated Dioxins and Related Compounds, Vienna, September 1993; and D. Cantor, G. Holder, W. Cantor, P. Kahn, G. Rodgers, G. Smoger, W. Swain, H. Berger, and S. Suffin, "*In Utero* and Postnatal Exposure to 2,3,7,8-TCDD in Times Beach, Missouri: 2. Impact on Neurophysiological Functioning," Dioxin '93.

"The EPA's reassessment . . ." refers to L. Gray and J. Ostby, "*In Utero* 2,3,7,8-Tetrachlorodibenzo-p-dioxin (TCDD) Alters Reproductive Morphology and Function in Female Rat Offspring," *Toxicology and Applied Pharmacology* (in press, 1995). Also see L. Gray, W. Kelce, E. Monosson, J. Ostby, and L. Birnbaum, "Exposure to TCDD During Development Permanently Alters Reproductive Function in Male Long Evans Rats and Hamsters: Reduced Ejaculated Epididymal Sperm Numbers and Sex Accessory Gland Weights in Offspring with Normal Androgenic Status," *Toxicology and Applied Pharmacology* 131:108–18 (1995).

"Well before their . . ." mentions the work of Dorothea Sager; see D. Sager, D. Girard, and D. Nelson, "Early Postnatal Exposure to PCBs: Sperm Function in Rats," *Environmental Toxicology and Chemistry* 10:737–46 (1991) for an overview.

CHAPTER 8: HERE, THERE, AND EVERYWHERE

Drs. Soto and Sonnenschein published a paper that describes how they arrived at their hypothesis concerning estrogenic action on cellular proliferation, "Mechanism of Estrogen Action on Cellular Proliferation: Evidence for Indirect and Negative Control on Cloned Breast Tumor Cells," *Biochemical and Biophysical Research Communication* 122:1097–1103 (1984). This was followed by A. Soto and C. Sonnenschein, "Cell Proliferation of Estrogen-Sensitive Cells: The Case for Negative Control," *Endocrine Reviews* 8:44–52 (1987).

"By 1985 . . ." describes work that is discussed more in depth in A. Soto and C. Sonnenschein, "The Role of Estrogens on the Proliferation of Human Breast Tumor Cells (MCF-7)," *Journal of Steroid Biochemistry* 23:87–94, (1985); and C. Sonnenschein, J. Papendorp, and A. Soto, "Estrogenic Effect of Tamoxifen and Its Derivatives on the Proliferation of MCF-7 Human Breast Tumor Cells," *Life Sciences* 37:387–94 (1985).

As a follow-up to the paragraph starting "In the end . . . ," see A. Soto, H. Justicia, J. Wray, and C. Sonnenschein, "p-Nonylphenol: A Estrogenic Xenobiotic Released from 'Modified' Polystyrene," *Environmental Health Perspectives* 92:167–73 (1991); and A. Soto, T. Lin, H. Justicia, R. Silvia, and C. Sonnenschein, "An 'In Culture' Bioassay to Assess the Estrogenicity of Xenobiotics (E-SCREEN)," in *Chemically Induced Alterations in Sexual and Functional Development: The Wildlife–Human Connection*, T. Colborn and C. Clement, eds., Princeton Scientific Publishing, 1992, pp. 295–309.

The passage starting "They also learned . . ." comes from a report by the Chemical Manufacturer Association's Alkylphenol and Ethoxylates Panel, "Alkylphenol Ethoxylates: Human Health and Environmental Effects," October 1993. For a recent discussion about the stability of the alkylphenols see W.-Y. Shiu, K.-C. Ma, D. Varhaníčková, and D. Mackay, "Chlorophenols and Alkylphenols: A Review and Correlation of Environmentally Relevant Properties and Fate in an Evaluative Environment," *Chemosphere* 29(6):1155–1224 (1994). These chemicals have little tendency to evaporate and therefore remain in water and soil. The authors admit that their work is "merely a first attempt to elucidate the environmental fate of this important and interesting class of chemicals."

"By strange coincidence . . ." refers to A. Krishnan, P. Stathis, S. Permuth, L. Tokes, and D. Feldman, "Bisphenol-A: An Estrogenic Substance Is Released from Polycarbonate Flasks During Autoclaving," *Endocrinology* 132(8):2279–86 (1993). Officials of GE Plastics Company contend that polycarbonate containers are unlikely to leach bisphenol-A in normal use because they would not be subjected to the high temperatures used in the Stanford lab for sterilization. (This is based on conversations with Diana Nichols, communications, and Tim Ullman, manager of global product stewardship, Pittsfield, Massachusetts.) The authors have not seen independent tests that assess the leaching properties of polycarbonates at various water temperatures in the presence of various kinds of cleaning agents or after extended use.

For more information concerning the section that begins "In this same period . . . ," see J. Sumpter and S. Jobling, "Vitellogenesis as a Bio-marker for Oestrogen Contamination of the Aquatic Environment," in *The Proceedings of the Estrogens in the Environment Conference*, Environmental Health Perspectives Supplements (in press, 1995); S. Jobling, T. Reynolds, R. White, M. Parker, and J. Sumpter, "A Variety of Environmentally Persis-tent Chemicals, Including some Phthalate Plasticizers, Are Weakly Estro-genic," *Environmental Health Perspectives* 103(6):582–87 (1995); C. Purdom, P. Hardiman, V. Bye, N. Eno, C. Tyler, and J. Sumpter, "Estrogenic Effects of Effluents from Sewage Treatment Works," Chemistry and Ecology 8:275–85 (1994); and S. Jobling and J. Sumpter, "Detergent Components in Sewage Effluent Are Weakly Oestrogenic to Fish: An *In Vitro* Study Using Rainbow Trout (*Oncorhynchus mykiss*) Hepatocytes," *Aquatic Toxicology* 27:361–72 (1993).

"Spurred by the Tufts . . ." refers to J. Brotons, M. Olea-Serrano, M. Vil lalobos, V. Pedraza, N. Olea, "Xenoestrogens Released from Lacquer Coat-ings in Food Cans," *Environmental Health Perspectives* 103(6):608–12 (1995).

"All of the incidents . . ." refers to W. Kelce, C. Stone, S. Laws, L. Gray, J. Kemppainen, and E. Wilson, "Persistent DDT Metabolite p,p'-DDE Is a Potent Androgen Receptor Antagonist," *Nature* 375:581–85 (1995). In their continuing work every estrogenic compound they have tested thus far also binds to the androgen (male) receptor and the proges-terone receptor. It appears that not only are the receptors promiscuous, but our hormones and their synthetic copycats are promiscuous as well (per-sonal communication with Earl Gray, July 1995).

"Such assertions seem . . ." is based on a discussion with an attorney, David Vladeck, director of the Public Citizen Litigation Group, Washington, D.C., who has litigated cases on trade secrets.

For more reading on global and regional contaminant trends (mentioned in the paragraph starting "Oftentimes, the studies . . ."), see B. Loganathan and K. Kannan, "Global Organochlorine Contamination Trends: An Overview," *Ambio* 23(3):187–91 (1994); and B. Loganathan, S. Tanabe, Y. Hidaka, M. Kawano, H. Hidaka, and R. Tatsukawa, "Temporal Trends of Persistent Organochlorine Residues in Human Adipose Tissue from Japan, 1928–1985," *Environmental Pollution* 81:31–39 (1993).

For historical information concerning the information on chemical production see A. Ihde, *The Development of Modern Chemistry*, Harper and Row, 1970; and M. Holdgate, A *Perspective of Environmental Pollution*, Cambridge University Press, 1979.

"Around the world . . ." uses figures from *The World Environment*

1972–1992: Two Decades of Challenge, M. Tolba and O. El-Kholy, eds., Chapman and Hall, 1992.

For more information on the production of pesticides, see C. Edwards "The Impact of Pesticides on the Environment," in *The Pesticide Question: Environment, Economics, and Ethics*, D. Pimentel and H. Lehman, eds., Chapman and Hall, 1993; D. Pimentel, "The Dimensions of the Pesticide Question," in *Ecology, Economics, Ethics: The Broken Circle*, F. Bormann and S. Kellert, eds., Yale University Press, 1991; A. Aspelin, "Pesticide Industry Sales and Usage, 1992 and 1993 Market Estimates Report," U.S. Environmental Protection Agency Report EPA/733/K-94/001, 1994; U.S. Food and Drug Administration, "Food and Drug Administration Pesticide Program Residues in Foods—1989," *Journal of the Association of Official Analytical Chemistry* 73:127A–146A (1990); U.S. Congressional Office of Technology Assessment, *Pesticide Residues in Food: Technologies for Detection*, Washington, D.C., 1988; S. Dogheim, E. Nasr, M. Almaz, and M. El Tohamy, "Pesticide Residues in Milk and Fish Samples Collected in Two Egyptian Governorato," *Journal of the Association of Official Analytical Chemistry* 73:19–21 (1990) [cited in Pimentel]; D. Acquay, M. Biltonen, P. Rice, M. Silva, J. Nelson, V. Lipner, S. Giordano, A. Horowitz, and M. D'Amore, "Assessment of Environmental and Economic Impacts of Pesticide Use," in *The Pesticide Question*; and B. Hileman, "Concerns Broaden over Chlorine and Chlorinated Hydrocarbons," *Chemical and Engineering News*, April 19, 1993, pp. 11–20.

The paragraph starting "The world trade . . ." closes with a statement supported by Loganathan et al., 1994.

"In 1991, the U.S. . . ." refers to the findings of Carl Smith at the Foundation for Advancements in Science and Education (FASE), Los Angeles, California, reported in "Exporting Banned and Hazardous Pesticides, 1991 Statistics: The Second Export Survey by the FASE Pesticide Project," 1993. Using customs records, Smith has since discovered that approximately a ton of DDT per day was shipped from the United States in 1992 (personal communication, July 1995). Smith also notes that customs records as a rule omit the technical names of pesticides; thus his figures are conservative estimates.

"Contrary to assertions . . ." refers to Kelce et al., 1995.

CHAPTER 9: CHRONICLE OF LOSS

For further reading on the status of the St. Lawrence beluga population, see L. Pippard, "Status of the St. Lawrence River Population of Beluga, *Delphinapterus leucas*," *The Canadian Field-Naturalist* 99:438–50 (1985).

The population figures cited in the paragraph starting "For years, scientists . . ." are based on aerial surveys in 1990, cited by P. Béland in a report for the Canadian World Wildlife Fund's Wildlife Toxicology Fund, "Toxicology and Pathology of Marine Mammals," May 1992. See also P. Béland, untitled article in *Whalewatcher: Journal of the American Cetacean Society* 28(1):3–5 (1994); and P. Béland, A. Vézina, and D. Martineau, "Potential for Growth of the St. Lawrence (Québec, Canada) Beluga Whale (*Delphinapterus leucas*) Population Based on Modelling," *Journal du Conseil international pour Exploration de la Mer* 45:22–32 (1988).

For further information after reading the paragraph starting "Even the first . . . ," see D. Martineau, P. Béland, C. Desjardins, and A. Lagacée, "Levels of Organochlorine Chemicals in Tissues of Beluga Whales (*Delphinapterus leucas*) from the St. Lawrence Estuary, Québec, Canada," *Archives of Environmental Contamination and Toxicology* 16:137–47 (1987); P. Béland, S. De Guise, C. Girard, A. Lagacée, D. Martineau, R. Michaud, D. Muir, R. Norstrom, E. Pelletier, S. Ray, and L. Shugart, "Toxic Compounds and Health and Reproductive Effects in St. Lawrence Beluga Whales," *Journal of Great Lakes Research* 19(4):766–75 (1993); and S. De Guise, A. Lagacée, and P. Béland, "Tumors in St. Lawrence Beluga Whales (*Delphinapterus leucas*)," *Veterinary Pathology* 31:444–49 (1994).

"The autopsy continued . . . ," which discusses Booly's condition, can be further explored in S. De Guise, A. Lagacée, and P. Béland, "True Hermaphroditism in a St. Lawrence Beluga Whale (*Delphinapterus leucas*)," *Journal of Wildlife Disease* 30(2):287–90 (1994).

"Was Booly an accident . . ." suggests that pollution may have reached a peak about the time Booly was born; see G. Sanders, S. Eisenreich, and K. Jones, "The Rise and Fall of PCBs: Time-Trend Data from Temperate Industrialized Countries," *Chemosphere* 29 (9–11):2201–2208 (1994). Sanders et al. provide figures for PCB loading in water and peat compared with production in the United Kingdom (Loch Ness) and the United States that support that conclusion.

"Such reproductive problems . . ." refers to P. Reijnders, "Reproductive Failure in Common Seals Feeding on Fish From Polluted Coastal Waters," *Nature* 324:456–57 (1986). For more reading about the immune status of marine mammals, see A. Osterhaus, "Seal Death," *Nature* 334:301–302 (1988); A. Osterhaus, J. Groen, P. De Vries, F. UytdeHaag, B. Klingeborn, and R. Zarnke, "Canine Distemper Virus in Seals," *Nature* 335:403–404 (1988); A. Osterhaus and E. Vedder, "Identification of Virus Causing Recent Seal Deaths," *Nature* 335:20 (1988); A. Osterhaus, J. Groen, F. Uytdehaag, I. Visser, M. Bildt, A. Bergman, and B. Klingeborn, "Distemper Virus in Baikal Seals," *Nature* 338:209–10, 1989; P. Ross, R. de Swart, I. Visser, L. Vedder, W. Murk, W. Bowen, and A. Osterhaus, "Relative Immunocompetence of the Newborn Harbour Seal, *Phoca vitulina*," *Veterinary Immunology and Immunopathology* 42:331–48 (1994); P. Ross, R. de Swart, P. Reijnders, H. Van Loveren, J. Vos, and A. Osterhaus, "Contaminant-Related Suppression of Delayed-Type Hypersensitivity and Antibody Responses in Harbor Seals Fed Herring from the Baltic Sea," Environmental Health Perspectives 103(2):162–67 (1995); and R. de Swart, "Impaired Immunity in Seals Exposed to Bioaccumulated Environmental Contaminants," Ph.D. thesis, Erasmus University, Rotterdam, Netherlands, 1995.

For a recent discussion about the St. Lawrence beluga whale population, see S. De Guise, D. Martineau, P. Béland, and M. Fournier, "Possible Mechanisms of Action of Environmental Contaminants on St. Lawrence Beluga Whales (*Delphinapterus leucas*)," *Environmental Health Perspectives Supplements* 103(4):73–77 (1995).

The close of the paragraph that opens "The first clue . . ." refers to M. Roelke, J. Martenson, and S. O'Brien, "The Consequences of Demographic Reduction and Genetic Depletion in the Endangered Florida Panther," *Current Biology* 3:340–50 (1993). See also C. Facemire, T. Gross, and L. Guillette, "Reproductive Impairment in the Florida Panther: Nature or Nurture?" *Environmental Health Perspectives Supplements* 103(4):79–86 (1995).

"With this insight . . ." refers to A. Woodward et al., 1993, cited in Chapter 1, and L. Guillette, T. Gross, G. Masson, J. Matter, H. Percival, and A. Woodward, "Developmental Abnormalities of the Gonad and Abnormal Sex Hormone Concentrations in Juvenile Alligators from Contaminated and Control Lakes in Florida," *Environmental Health Perspectives* 102:680–88, 1994; L. Guillette, D. Crain, A. Rooney, and D. Pickford, "Organization Versus Activation: The Role of Endocrine-Disrupting Contaminants (EDCs) During Embryonic Development in Wildlife," *Environmental Health Perspectives Supplement* (in press); and L. Guillette and D. Crain, "Endocrine-Disrupting Contaminants and Reproductive Abnormalities in Reptiles," *Comments on Toxicology* (in press).

For more information about temperature influence (mentioned in the paragraph starting "In turtles, sex is determined . . ."), see D. Crews, J. Bergeron, J. Bull, D. Flores, A. Tousignant, J. Skipper, and T. Wibbels, "Temperature-Dependent Sex Determination in Reptiles: Proximate Mechanisms, Ultimate Outcomes and Functional Outcomes," *Developmental Genetics* 15:297–312 (1994). This paper describes the authors' findings following the painting of turtle eggs with single PCB congeners during incubation.

"In the Great Lakes . . ." refers to the doctoral dissertation of W. Bowerman, "Regulation of Bald Eagle (*Haliaeetus leucocephalus*) Productivity in the Great Lakes Basin: An Ecological and Toxicological Approach," Michigan State University, Department of Fisheries and Wildlife, 1993, and personal communication with Bowerman, and with Dave Best and Letha Williams, U.S. Fish and Wildlife Service. For an excellent review and analysis of the literature on contaminants in Great Lakes birds, see J. Giesy, J. Ludwig, and D. Tillitt, "Deformities in Birds of the Great Lakes Region: Assigning Causality," *Environmental Science and Technology* 28(3):128–35 (1994).

"An eagle's diet . . ." describes the research of Karen Kozie, as reported in T. Colborn, "Epidemiology of Great Lakes Bald Eagles," *Journal of Toxicology and Environmental Health* 33(4):395–453 (1991).

"Based on environmental . . ." refers to the syndrome called GLEMEDS cited in Chapter 1.

"In 1993, the bald eagles . . ." refers to findings in W. Bowerman, T. Kubiak, J. Holt, D. Evans, R. Eckstein, C. Sindelar, D. Best, and K. Kozie, "Observed Abnormalities in Mandibles of Nestling Bald Eagles Haliaeetus leucocephalus," *Bulletin of Environmental Contamination and Toxicology* 53:450–57 (1994), and personal communication with Carol Schuler, Portland Field Office, U.S. Fish and Wildlife Service.

The section of this chapter devoted to the mink is taken from the writings of Richard Aulerich and Robert Ringer as cited in R. Aulerich et al., 1973, mentioned in the notes to Chapter 1, and R. Aulerich and R. Ringer, "Current Status of PCB Toxicity to Mink, and Effect on Their Reproduction," *Archives of Environmental Contamination and Toxicology* 6:279–92 (1977).

The section of this chapter devoted to otters (starting with "In Britain and Europe . . .") leans on the following literature: C. Mason, "Role of Contaminants in the Decline of the European Otter," *Proceeding of the Expert Consultation Meeting on Mink and Otter*, International Joint Commission, Windsor, Ontario, 1991; R. Foley, S. Jackling, R. Sloan, and M. Brown, "Organochlorine and Mercury Residues in Wild Mink and Otter: Comparison with Fish," *Environmental Toxicology and Chemistry* 7:363–74 (1988); and C. Henny, L. Blus, S. Gregory, and C. Stafford, "PCBs and Organochlorine Pesticides in Wild Mink and River Otters from Oregon," in *Proceedings of Worldwide Furbearer Conference*, J. Chapman and D. Pursley, eds., Frostburg, Maryland, 1981, pp. 1763–80.

The history discussed in the paragraph starting "The demise of . . ." is described in much greater detail in T. Colborn, A. Davidson, S. Green, R. Hodge, C. Jackson, and R. Liroff, *Great Lakes, Great Legacy?*, The Conservation Foundation and the Institute for Research on Public Policy, 1990. The reference to dioxin in this paragraph is based on M. Walker and R. Peterson, "Toxicity of Polychlorinated Dibenzo-p-Dioxins, Dibenzofurans, and Biphenyls During Early Development in Fish," in *Chemically Induced Alterations in Sexual and Functional Development: The Wildlife-Human Connection*, T. Colborn and C. Clement, eds., Princeton Scientific Publishing, 1992, pp.195–202; and P. Cook, D. Kuehl, M. Walker, and R. Peterson, "Bioaccumulation and Toxicity of TCDD and Related Compounds in Aquatic Ecosystems," *Banbury Report 35: Biological Basis for Risk Assessment of Dioxins and Related Compounds*, Cold Spring Harbor Laboratory Press, 1991, pp. 143–67; and personal communication with Phil Cooke, U.S. Environmental Protection Agency, Duluth, Minnesota, 1995.

For more information about the conditions described in the passage starting "Egg mortality . . . ," see J. Leatherland, "Endocrine and Reproductive Function in Great Lakes Salmon," in *Chemically Induced Alterations*, pp. 129–45.

The paragraph commencing "In Europe and Scandinavia . . ." leads into a section on marine mammals that is based on work already cited throughout. See especially the section in Chapter 1 on marine mammal die-offs. For further reviews, see P. Reijnders and S. Brasseur, "Xenobiotic Induced Hormonal and Associated Developmental Disorders in Marine Organisms and Related Effects in Humans: An Overview," in *Chemically Induced Alterations*, pp. 159–74; J. Raloff, "Something's Fishy: Marine Epidemics May Signal Environmental Threats to the Immune System," *Science News*, 146:8–9 (1994); and T. Colborn and M. Smolen, "An Epidemiological Analysis of Persistent Organochlorine Contaminants in Cetaceans," *Reviews of Environmental Contaminants and Toxicology* (in press).

"A second study . . ." refers to the work of G. Lahvis and coworkers R. Wells, D. Casper, and C. Via, reported in *"In Vitro* Lymphocyte Response of Bottlenose Dolphins (*Tursiops truncatus*): Mitogen-Induced Proliferation," *Marine Environmental Research* 35:115–19 (1993); and G. Lahvis, R. Wells, D. Kuehl, J. Stewart, H. Rhinehart, and C. Via, "Decreased Lymphocyte Responses in Free-Ranging Bottlenose Dolphins (*Tursiops truncatus*) Are Associated with Increased Concentrations of PCBs and DDT in Peripheral Blood," *Environmental Health Perspectives Supplements* 103(4):67–72 (1995).

As a follow-up to the paragraph starting "Even less is known . . . ," see R. Stebbins and N. Cohen, *A Natural History of Amphibians*, Princeton University Press, 1995; and A. Blaustein, D. Wake, and W. Sousa, "Amphibian Declines: Judging Stability, Persistence, and Susceptibility of Populations to Local and Global Extinctions," *Conservation Biology* 8(1):60–71 (1994).

CHAPTER 10: ALTERED DESTINIES

The paragraph opening "The relevance of . . ." leads into a discussion of the work of F. vom Saal, S. Nagel, P. Palanza, M. Boechler, S. Parmigiani, and W. Welshons, reported in "Estrogenic Pesticides: Binding Relative to Estradiol in MCF-7 Cells and Effects of Exposure During Fetal Life on Subsequent Territorial Behavior in Male Mice," *Toxicology Letter* (in press, 1995).

"At the end . . ." refers to the Wingspread Consensus Statement, which is Chapter One in *Chemically Induced Alterations in Sexual and Functional Development: The Wildlife-Human Connection*, T. Colborn and C. Clement, eds., Princeton Scientific Publishing, 1992. It is printed in the Appendix of this book.

"Based on warnings . . ." is supported by R. Newbold, B. Bullock, and J. Mclachlan, "Uterine Adenocarcinoma in Mice Following Developmental Treatment with Estrogens: A Model for Hormonal Carcinogenesis," *Cancer Research* 50:7677–81 (1990); H. Bern and F. Talamantes, "Neonatal Mouse Models and their Relation to Disease in the Human Female," in *Developmental Effects of Diethylstilbestrol (DES) in Pregnancy*, A. Herbst and H. Bern, eds., Thieme-Stratton, 1981, pp. 129–47; and B. Bullock, R. Newbold, and J. McLachlan, "Lesions of Testis and Epididymis Associated with Prenatal Diethylstilbestrol," *Environmental Health Perspectives* 77:29–31 (1988).

"Laboratory experiments . . ." refers to L. Birnbaum, "Endocrine Effects of Prenatal Exposures to PCBs, Dioxins, and Other Xenobiotics: Implications for Policy and Future Research," *Environmental Health Perspectives* 102(8):676–79 (1994).

"The most dramatic . . ." refers to E. Carlsen et al., 1992, cited in Chapter 1.

The passage starting "The study is . . ." points out the difficulty of recognizing a reduction in sperm count. Traditionally, andrologists considered a sperm count under 50 million sperm per milliliter a signal of a reduction in fertility. More recently, this benchmark has been lowered to 20 million sperm per milliliter. However, even at 20 million sperm per milliliter most men can produce viable offspring, given sufficient opportunity. The skeptics include P. Bromwich, J. Cohen, I. Stewart, and A. Walker, "Decline in Sperm Counts: An Artifact of Changed Reference Range of 'Normal'?" *British Medical Journal* 309:19–22 (1994); and G. Olsen, K. Bodner, J. Ramlow, C. Ross, and L. Lipshultz, "Have Sperm Counts Been Reduced 50 Percent in 50 Years? A Statistical Model Revisited," *Fertility and Sterility* 63(4):887–93 (1995).

"This debate stimulated . . ." refers to J. Auger, J. Kunstmann, F. Czyglik, and P. Jouannet, "Decline in Semen Quality Among Fertile Men in Paris During the Past 20 Years," *New England Journal of Medicine* 332(5):281–85 (1995); K. Van Waeleghem, N. De Clercq, L. Vermeulen, F. Schoonjans, and F. Comhaire, "Deterioration of Sperm Quality in Young Belgian Men During Recent Decades," in *Abstracts of the Annual Meeting of the ESHRE,* Brussels, 1994, p. 73; and D. Irvine, "Falling Sperm Quality," letter to the editor, *British Medical Journal,* 309:131 (1994).

"When Sharpe and Skakkebaek . . ." refers to their article, R. Sharpe and N. Skakkebaek, "Are Oestrogens Involved in Falling Sperm Counts and Disorders of the Male Reproductive Tract?" *Lancet* 341:1392–95 (1993). For more information on undescended testicles, see C. Chilvers, M. Pike, D. Forman, K. Fogelman, and M. Wadsworth, "Apparent Doubling of Frequency of Undescended Testis in England and Wales in 1962–81," *Lancet* 330–32 (1984); and J. Hutson, M. Baker, M. Terada, B. Zhou, and G. Paxton, "Hormonal Control of Testicular Descent and the Cause of Cryptorchidism," *Reproduction, Fertility, and Development* 6:151–56 (1994). The latter describes what is known about testicular descent, the multifactorial etiology of cryptorchidism, and mentions the role estrogens, anti-androgens, and Müllerian inhibiting substances (MIS) play in prenatal differentiation of tissue that could become ovary or testis.

"Synthetic chemicals have . . ." mentions Dorothea Sager; her work was cited in Chapter 7. The paragraph closes with a description of the work of Brian Bush and associates cited in Chapter 6.

"Before biological research . . ." refers to reports from the National Center for Health Statistics, Hyattsville, Maryland; and Congressional Office of Technology Assessment, "Infertility: Medical and Social Choices," U.S. Government Printing Office, May 1988. The $2 billion is a figure gleaned from an interview with Joyce Zeitz of the American Fertility Society, Birmingham, Alabama.

"Prenatal exposure . . ." refers to a discussion in Chapter 3. See citations there.

"In ongoing studies . . ." touches on F. vom Saal, M. Dhar, V. Ganjam, and W. Welshons, "Prostate Hyperplasia and Increased Androgen Receptors in Adulthood Induced by Fetal Exposure to Estradiol in Mice," Abstract for the Fall Meeting of the Society for Basic Urologic Research, Stanford University, 1994.

"Even in males . . ." refers to S. Ho and M. Yu, "Selective Increase in Type II Estrogen-Binding Sites in the Dysplastic Dorsolateral Prostates of Noble Rats," *Cancer Research* 53:528–32 (1993), personal communication with Ho, 1995, and from a discussion with Dr. Ronald McDaniels, Abbott Pharmaceuticals, and a paper, J. Oesterling, "Benign Prostatic Hyperplasia: Diagnosis and Treatment Options," *New England Journal of Medicine* 332(2):99 (1995). Oesterling states that 400,000 transurethral resections of the prostate are performed each year at a cost of 5 billion dollars. Mc-Daniels estimates an additional 1 billion dollars for drug therapy.

The figures concerning breast cancer in the passage starting "Over the past . . ." come from a review of the National Cancer Institute, SEER Statistics, and reviews of existing data by the authors.

"Despite more sophisticated . . ." uses data from Centers for Disease Control, "Radical Prostatectomies—Wisconsin, 1982–1992," *Morbidity and Mortality Weekly Report* 42(32):620–21, 627 (1993); and "Trends in Prostate Cancer—United States, 1980–1988," *Morbidity and Mortality Weekly Report* 41:401–404 (1992).

"In tubal or ectopic . . ." states that repeated tubal pregnancies can result in infertility. There are no official estimates on the proportion of female infertility caused by endometriosis, but Zeitz of the American Fertility Society estimates that it might be responsible for twenty percent of the cases. The other two leading causes are ovulatory and tubal problems—problems reported in DES daughters.

For historical information about "Endometriosis, a poorly . . . ," see J. Older, "Leaches and Laudanum: Grandmother and You: Historical Highlights," *Endometriosis*, Scribners, 1964. The National Institute of Child Health publication mentioned is "Facts About Endometriosis," No. 91-2413 (no date).

"After years of debate . . ." was drawn from a conversation with Mary Lou Balweg, The Endometriosis Association, Milwaukee, Wis. 1994; and from S. Rier, D. Martin, R. Bowman, W. Dmowski, and J. Becker, "Endometriosis in Rhesus Monkeys (*Macaca mulatta*) Following Chronic Exposure to 2,3,7,8-Tetrachlorodibenzo-p-dioxin," *Fundamental and Applied Toxicology* 21:433–41 (1993).

"Animal studies also . . ." also leans on Rier et al., 1993, and mentions V. Leoni, L. Fabiani, G. Marinelli, G. Puccetti, G. Tarsitani, A. De Carolis, N. Vescia, A. Morini, V. Aleandri, V. Pozzi, F. Cappa, and D. Barbati, "PCB and Other Organochlorine Compounds in Blood of Women with or Without Miscarriage: A Hypothesis of Correlation," *Ecotoxicology and Environmental Safety* 17:1–11 (1989).

In connection with the paragraph opening "But by far . . . ," see D. Davis and H. Freeman, "An Ounce of Prevention," *Scientific American*, September 1994, p. 112.

"Because total estrogen . . ." refers to D. Hunter and K. Kelsey, "Pesticide Residues and Breast Cancer: The Harvest of a Silent Spring?" *Journal of the National Cancer Institute* 85(8):598–99 (1993).

For more information concerning the paragraph starting "The most notable . . . ," see P. Pujol, S. Hilsenbeck, G. Chamness, and R. Elledge, "Rising Levels of Estrogen Receptor in Breast Cancer over 2 Decades," *Cancer* 74(5):1601–1606 (1994).

For the passage that begins "In 1993, a group . . . ," see D. Davis, H. Bradlow, M. Wolff, T. Woodruff, D. Hoel, and H. Anton-Culver, "Medical Hypothesis: Xenoestrogens as Preventable Causes of Breast Cancer," *Environmental Health Perspectives* 101(5):372–77 (1993).

"Researchers have been . . ." refers to É. Dewailly, S. Dodin, R. Verreault, P. Ayotte, L. Sauvé, J. Morin, and J. Brisson, "High Organochlorine Body Burden in Women with Estrogen Receptor-Positive Breast Cancer," *Journal of the National Cancer Institute* 86(3):232–34 (1994).

"Two other studies . . ." refers to M. Wolff, P. Toniolo, E. Lee, M. Rivera, and N. Dubin, "Blood Levels of Organochlorine Residues and Risk of Breast Cancer," *Journal of the National Cancer Institute* 85(8):648–52 (1993). However, N. Krieger, M. Wolff, R. Hiatt, M. Rivera, J. Vogelman, and N. Orentreich, "Breast Cancer and Serum Organochlorines: A Prospective Study Among White, Black, and Asian Women," *Journal of the National Cancer Institute* 86(8):589–99 (1994), found that breast cancer patterns varied among racial groups. They found a positive but not statistically significant association between DDT and breast cancer in Caucasians and African–Americans, but no association for Asians. See a letter from M. Wolff and P. Landrigan, "Response to Environmental Estrogens Stir Debate," *Science* 266:526–27 (1994).

"Among cancer victims . . ." refers to D. Hoel, D. Davis, A. Miller, E. Sondik, and A. Swerdlow, "Trends in Cancer Mortality in 15 Industrialized Countries, 1969–1986," *Journal of the National Cancer Institute* 84(5):313–20 (1992).

"What little we . . ." refers to *The Merck Manual of Diagnosis and Therapy*, 15th ed., R. Berkow and A. Fletcher, eds., Merck Sharp and Dohme Research Laboratories, 1987, p. 1978. For more information on attention deficit disorder, see P. Hauser, A. Zametkin, P. Martinez, B. Vitiello, J. Matochik, A. Mixson, and B. Weinstraub, "Attention Deficit-Hyperactivity Disorder in People with Generalized Resistance to Thyroid Hormone," *New England Journal of Medicine* 328(14):997–1001 (1993).

"It has long . . ." mentions S. Porterfield, "Vulnerability of the Developing Brain to Thyroid Abnormalities: Environmental Insults to the Thyroid System," *Environmental Health Perspectives* 102(2):125–30 (1994);

S. Porterfield and S. Stein, "Thyroid Hormones and Neurological Development: Update 1994," *Endocrine Reviews* 3(1):357–63 (1994); and S. Porterfield and C. Hendrich, "The Role of Thyroid Hormones in Prenatal and Neonatal Neurological Development—Current Perspectives," *Endocrine Reviews* 14(1):94–106 (1993).

For more information concerning the paragraph starting "PCBs and dioxin . . . ," see H. Pluim, J. de Vijlder, K. Olie, J. Kok, T. Vulsma, D. van Tijn, J. van der Slikke, and J. Koppe, "Effects of Pre- and Postnatal Exposure to Chlorinated Dioxins and Furans on Human Neonatal Thyroid Hormone Concentrations," *Environmental Health Perspectives* 101(6):504–508 (1993); D. Bombick, J. Jankun, K. Tullis, and F. Matsumura, "2,3,7,8-Tetrachlorodibenzo p dioxin Causes Increases in Expression of c-erb-A and Levels of Protein-Tyrosine Kinases in Selected Tissues of Responsive Mouse Strains," *Proceedings of the National Academy of Sciences* 85:4128–32 (1988); A. Brouwer, "Inhibition of Thyroid Hormone Transport in Plasma of Rats by Polychlorinated Biphenyls," *Archives of Toxicology Supplements* 13:440 45 (1989); A. Brouwer, F. Klasson-Wehler, M. Bokdam, D. Morse, and W. Traag, "Competitive Inhibition of Thyroxin Binding to Transthyretin by Monohydroxy Metabolites of 3,4,3',4'-Tetrachlorobiphenyl," *Chemosphere* 20(7–9):1257–62 (1990); D. Morse, H. Koeter, A. Smits van Prooijen, and A. Brouwer, "Interference of Polychlorinated Biphenyls in Thyroid Hormone Metabolism: Possible Neurotoxic Consequences in Fetal and Neonatal Rats," *Chemosphere* 25(1–2):165–68 (1992); and D. Ness et al., 1993, cited in Chapter 6.

"Based on the . . ." refers to a conversation with Linda Birnbaum, 1994; and H. Tilson, J. Jacobson, and W. Rogan, "Polychlorinated Biphenyls and the Developing Nervous System: Cross-Species Comparisons," *Neurotoxicology and Teratology* 12:239–48 (1990). This paper presents an excellent review. The studies that Tilson and coauthors referred to did not permit the researchers to separate the effects of exposure in the womb from exposure via breast milk. However, the researchers later determined that the effects they were measuring were the result of prenatal exposure.

"Much of our . . ." cites Y. Guo, G. Lambert, and C.-C. Hsu, "Growth Abnormalities in the Population Exposed to PCBs and Dibenzofurans," paper presented at Children's Environmental Health Network Conference, 1994. See also M. Yu, C. Hsu, Y. Guo, T. Lai, S. Chen, and J. Luo, "Disordered Behavior in the Early-Born Taiwan Yucheng Children," *Chemosphere* 29(9–11):2413–22 (1994); for further reading about the rice oil incidents, see W. Rogan, B. Gladen, K. Hung, S. Koong, L. Shih, J. Taylor, Y. Wu,

D. Yang, N. Ragan, and C.-C. Hsu, "Congenital Poisoning by Polychlorinated Biphenyls and Their Contaminants in Taiwan," *Science* 241:334–36, (1988); and W. Rogan, "PCBs and Cola-Colored Babies: Japan, 1968, and Taiwan, 1979," *Teratology* 26:259–61 (1982).

"The first study . . ." refers to J. Jacobson, S. Jacobson, P. Schwartz, G. Fein, and J. Dowler, "Prenatal Exposure to an Environmental Toxin: A Test of the Multiple Effects Model," *Developmental Psychology* 20(4):523–32 (1984). There were 242 infants of fisheaters and 71 infants whose mothers ate no Lake Michigan fish in the study. The figure on the value of the sports fishing industry in the Great Lakes was taken from D. Talheim, "Economics of Great Lakes Fisheries: A 1985 Assessment," Special Economic Report, Great Lakes Fishery Commission, Ann Arbor, Mich., 1987.

"The second study . . ." refers to B. Gladen, W. Rogan, P. Hardy, J. Thullen, J. Tingelstad, and M. Tully, "Development After Exposure to Polychlorinated Biphenyls and Dichlorodiphenyl Dichloroethene Transplacentally and Through Human Milk," *Journal of Pediatrics* 113(6):991–95 (1988); and W. Rogan, B. Gladen, J. McKinney, N. Carreras, P. Hardy, J. Thullen, J. Tinglestad, and M. Tully, "Neonatal Effects of Transplacental Exposure to PCBs and DDE," *Journal of Pediatrics* 109(2):335–41 (1986).

For more extensive information about the section that begins "At the State University . . . ," see H. Daly, "The Evaluation of Behavioral Changes Produced by Consumption of Environmentally Contaminated Fish," in *The Vulnerable Brain and Environmental Risks: Vol. 1. Malnutrition and Hazard Assessment*, Chapter 7, R. Isaacson and K. Jensen, eds., Plenum, 1992, pp. 151–71; H. Daly, "Laboratory Rat Experiments Show Consumption of Lake Ontario Salmon Causes Behavioral Changes: Support for Wildlife and Human Research Results," *Journal of Great Lakes Research* 19(4):784–88 (1993); and H. Daly, "Reward Reductions Found More Aversive by Rats Fed Environmentally Contaminated Salmon," *Neurotoxicology and Teratology* 13:449–53 (1991).

"In May 1995 . . ." refers to a paper presented by H. Daly at the 38th Annual Conference of the International Association for Great Lakes Research (IAGLR), East Lansing, Michigan, and a May 27, 1995, press release by the State University of New York, Oswego, pointing out that this "is the first large-scale replication and extension of the Jacobsons' Lake Michigan Infant Cohort Study." Only associations with the amount of fish eaten over a lifetime were reported at this time. The statistical results of chemical analysis on mother's blood, placental blood, and breast milk will be released in a series of future papers.

"In 1991, Dr. Simon LeVay . . ." refers to S. LeVay, "A Difference in

Hypothalamic Structure Between Heterosexual and Homosexual Men," *Science* 253:1034–37, 1991; and S. LeVay, *The Sexual Brain*, MIT Press, 1993, p. 168.

Upon finishing this chapter the reader might want to see "Male Reproductive Health and Environmental Oestrogens," editorial in *Lancet* 345:933–35 (1995); or R. Sharpe, "Another DDT Connection," News and Views article in *Nature* 375:538–39 (1995); and, as well, "Masculinity at Risk," Opinion article in *Nature* 375:522 (1995). See also A. Abell, E. Ernst, and J. Bonde, "High Sperm Density Among Members of Organic Farmers' Association," *Lancet* 343:1498 (1994).

CHAPTER 11: BEYOND CANCER

The first paragraph refers to H. Burlington and V. Lindeman, 1950, cited in Chapter 5.

"The study provided . . ." leans on T. Dunlap, *DDT: Scientists, Citizens, and Public Policy*, Princeton University Press, 1981, p. 70; and C. Bosso, *Pesticides and Politics: The Life Cycle of a Public Issue*, University of Pittsburgh Press, 1987.

"By the 1960s . . ." refers to R. Carson, *Silent Spring*, Houghton Mifflin, 1962.

For interesting reading about the 1972 decision to restrict the use of DDT (as mentioned in the paragraph that opens "Cancer holds a special . . ."), see U.S. Environmental Protection Agency I. F. and R. Docket 63, "Consolidated DDT Hearings: Opinion and Order of the Administrator," *Federal Register* 37 (131):13369–76, July 7, 1972.

"The federal appeals . . ." refers to *Environmental Defense Fund v. U.S. Department of Health Education and Welfare*, 1970, U.S. Court of Appeals, District of Columbia Circuit.

"Some critics of EPA . . ." refers to those who insist that low-dose exposure is tolerable; for example, P. Abelson, "Risk Assessments of Low-Level Exposures," editorial in *Science* 265:1507 (1994).

"Those studying . . ." mentions the complexity of feedback loops in the endocrine system; for example, within one low range of exposure one response may prevail—but as doses increase another response may set in, with part of the impact being to shut down the first response.

"As EPA toxicologist Linda Birnbaum . . ." refers again to her 1994 paper cited in Chapter 10. The Seveso studies by Bertazzi et al. were cited in Chapter 7. An earlier study looked for birth defects in children and concluded that "although the data collected failed to demonstrate any increased risk of birth defects associated with 2,3,7,8-tetrachlorodibenzo-p-dioxin, the number of exposed pregnancies was not big enough to show a low and specific teratogenic risk increase" (P. Mastroiacovo, A. Spagnolo, E. Marni, L. Meazza, R. Bertollini, and G. Segni, "Birth Defects in the Seveso Area After TCDD Contamination," *Journal of the American Medical Association* 259(11):1668–72 [1988]).

"If we are . . ." gets into a discussion about the weight-of-evidence approach to deal with difficult decisions concerning cause and effect. See G. Fox, "Scientific Principles in Applying the Weight of Evidence" in *Applying Weight of Evidence: Issues and Practice*, Canadian and U.S. International Joint Commission, June 1994; and G. Fox, "Practical Causal Inference for Ecoepidemiologists," *Journal of Toxicology and Environmental Health* 33:359–73 (1991).

Chapter 12: DEFENDING OURSELVES

"Unfortunately, however, solutions . . ." refers to findings like that of S. Hooper et al., 1990, cited in Chapter 6. For the closing sentence of this paragraph, see D. Patterson, Jr., G. Todd, W. Turner, V. Maggio, L. Alexander, and L. Needham, "Levels of Non-ortho-substituted (Coplanar), Mono- and Di-ortho-substituted Polychlorinated Biphenyls, Dibenzo-p-dioxins, and Dibenzofurans in Human Serum and Adipose Tissue," *Environmental Health Perspectives Supplements* 102(1):195–204 (1994); and once again see L. Birnbaum, 1994, cited in Chapter 10.

The following are only a few organizations that are structured to deal with the vast questions that are emerging as more and more people want to learn how to deal with synthetic chemicals in their environment:

- Rachel Carson Council, 8940 Jones Mill Road, Chevy Chase, MD 20815; tel. 301-652-1877; fax 301-951-7179. Provides information regarding pesticides and alternatives.
- National Coalition Against the Misuse of Pesticides (NCAMP), 701 E Street, SE, Washington, DC 20003; tel. 202-541-5450; fax 202-543-4791. Provides information regarding pesticide use.
- American PIE (American Public Information on the Environment), 31 North Main Street, P.O. Box 460, Marlborough, CT 06447-0460; tel. 800-320-APIE[2743]. Provides information about environmentally safe practices for home, office, and the outdoors. It can advise on

topics ranging from drinking water safety to household waste to lawn and garden practices.

- Enviro-Health, 100 Capitola Drive, Suite 108, Durham, NC 27713; tel. 800-NIEHS-94[643-4794]; fax 919-361-9408. This is the environmental health effects clearinghouse of the National Institute of Environmental Health Sciences (NIEHS), U.S. Department of Health and Human Services. Topics include human health, general health, chemical exposures, multiple chemical sensitivity, immunotoxicity, water and air quality, NIEHS research, education and information, hazardous waste, and more.
- Pesticide Action Network (North America), 116 New Montgomery Street, #810, San Francisco, CA 94105; tel. 415-541-9140; fax 415-541-9253; gopher site: gopher.econet.ape.org provides information on the global use of pesticides.

For further reading related to the section titled "Choose Your Food Intelligently," we recommend H. Needleman and P. Landrigan, *Raising Healthy Children*, Farrar, Straus, and Giroux, 1991, and A. Garland, *The Way We Grow*, Berkley, 1993.

"Clean fish are . . ." refers to health advisories for fish. The first U.S. review of fish advisories was D. Zeitlin, "State-Issued Fish Consumption Advisories: A National Perspective," for the National Oceanic and Atmospheric Administration, National Ocean Pollution Program Office, Washington, D.C., 1990.

"Avoid animal fat . . ." leads into a discussion about persistent chemicals building up in animal tissue. More and more laboratories are acquiring the equipment and technology to detect "low levels" of chemicals in food products, and as a result, new papers are appearing regularly in the scientific literature. Surprisingly, it appears as though dioxin is being delivered to millions daily through their favorite food, the hamburger.

An article that provides more food for thought concerning the paragraph starting "Researchers are only . . ." is K. Uvnäs-Moberg, "The Gastrointestinal Tract in Growth and Reproduction," *Scientific American*, July 1989, pp. 78–83. The author states that the gastrointestinal tract "is the largest endocrine gland in the body, and it has a significant role in the readjustment of metabolism that accompanies pregnancy as well as fetal and infant growth." In this and other writings the author notes that oxytocin levels increase in women during pregnancy and breast feeding and thus increase their social sensitivity and bonding. She also notes that an infant's weight doubles during its first six months. A six-week-old baby, weighing about 9 pounds, drinks approximately 22 ounces of milk a day. If the mother, weighing approximately 140 pounds, consumed a proportional amount of milk, she would drink about 10 quarts or 342 ounces a day. We should like to point out that this source of high energy for the infant also provides the medium to transport high doses of fat-soluble chemicals to the baby.

For more reading following the discussion in the paragraph starting "Never assume a pesticide . . . ," see C. Clement and T. Colborn, "Herbicides and Fungicides: A Perspective on Potential Human Exposure," in *Chemically Induced Alterations in Sexual and Functional Development: The Wildlife-Human Connection*, T. Colborn and C. Clement, eds., Princeton Scientific Publishing, 1992, pp. 347–64. The chapter cited reveals how little is known about the pesticide (Dacthal or DCPA) most frequently found in groundwater in the United States. Dacthal was discovered in trout and whitefish tissue from Siskiwit Lake on Isle Royale in Lake Superior, evidence that this product volatilizes and moves on air currents as well. See D. Swackhamer and R. Hites, "Occurrence and Bioaccumulation of Organochlorine Compounds in Fishes from Siskiwit Lake, Isle Royale, Lake Superior," *Environmental Science and Technology* 22(5):543–48 (1988).

"Don't be blasé . . ." discusses "inerts." See a video narrated by Dr. Mary O'Brien, "Inert Alert: Secret Poisons in Pesticides," The Northwest Coalition for Alternatives to Pesticides, 1991; available from S. Genasci, P.O. Box 1393, Eugene, OR 97440. See also D. Krewski, J. Wargo, and R. Rizek, "Risks of Dietary Exposure to Pesticides in Infants and Children," in *Monitoring Dietary Intakes*, I. Macdonald, ed., Springer-Verlag, 1991, pp. 75–89. The authors point out that nearly three hundred pesticide active ingredients used in commercial agricultural products are tolerated as food residues under U.S. EPA regulations.

"Be aware that . . ." cites the results of a report by the Attorney General of New York State, "Toxic Fairways: Risking Groundwater Contamination from Pesticides on Long Island Golf Courses," 1991.

For further reading about alternative production processes, we recommend P. Hawken, *The Ecology of Commerce*, Harper Business, 1993.

"Garden designers . . ." discusses redesigning yards and lawns to reduce chemical use. Once you have established a chemical-free yard, send to American PIE for a set of lawn flags boasting that your lawn is "SAFE TO PLAY ON." American PIE's address and phone number are given above in the citations for this chapter.

CHAPTER 13: LOOMINGS

For the closing section of the paragraph starting "If currently regulated . . . ,"
see V. Bhatnagar, J. Patel, M. Variya, K. Venkaiah, M. Shah, and S. Kashyap,
"Levels of Organochlorine Insecticides in Human Blood from Ahmedabad
(Rural), India," *Bulletin of Environmental Contamination and Toxicology*
48:302–307 (1992).

The decline discussed in the paragraph starting "Why did the Scholastic . . ." was described in a report of the Advisory Panel on the Scholastic Aptitude Test Score Decline for the College Entrance Examination Board, New
York, 1977.

"Consider, however . . ." refers to B. Weiss, "The Scope and Promise
of Behavioral Toxicology," in *Behavioral Measures of Neurotoxicity: Report
of a Symposium*, R. Russell, P. Flattau, and A. Pope, eds., National Research
Council, National Academy Press, 1990, pp. 395–413. The numbers of individuals affected at the high and low ends of the IQ scale are based on the
assumption that the standard deviation of 15 holds after a 5-point popula-
tion-wide IQ decrease.

"Other studies suggest . . ." discusses the research of Fred vom Saal
and coworkers at the University of Missouri cited in earlier chapters and the
research of Dr. Warren Porter and coworkers, Zoology Department, University of Wisconsin, Madison, in which they gave rats and mice free access
to water containing mixtures of pesticides and nutrients commonly found
in wells in Dane County; see W. Porter, S. Green, N. Debbink, and I. Carl-
son, "Groundwater Pesticides: Interactive Effects of Low Concentrations of
Carbamates, Aldicarb and Methomyl, and the Triazine Metribuzin on Thy-
roxine and Somatotropin Levels in White Rats," *Journal of Toxicology and
Environmental Health* 40:15–34 (1993).

CHAPTER 14: FLYING BLIND

The historical account of Midgely's work and other aspects of the ozone history leans on S. Cagin and P. Dray, *Between Earth and Sky: How CFCs
Changed Our World and Endangered the Ozone Layer*, Pantheon, 1993. We
also supplemented with our own research.

"Du Pont was . . ." mentions S. Rowland and M. Molina, "Stratospheric
Sink for Chlorofluoromethanes: Chlorine Atom-Catalyzed Destruction of
Ozone," *Nature*, June 28, 1974. The authors are recent Nobel Prize winners.

To our knowledge, endocrine disruption is not among the adverse health effects associated with reduced ozone in the stratosphere.

"The situation confronting . . ." cites a figure taken from a seminar presented by Brad Leinhart, Chlorine Chemical Council, Chemical Manufacturer's Association, Massachusetts Institute of Technology, January 26, 1994.

EPILOGUE

"In May 1996, a group of international experts . . ." refers to the Erice Statement, officially titled "Statement from the Work Session on Environmental Endocrine-Disrupting Chemicals: Neural, Endocrine, and Behavioral Effects" released in May 1996. This consensus statement and the scientific papers from this meeting is now in preparation and will be published by Princeton Scientific Publishing Co. in *Toxicology and Industrial Health: An International Journal.* For a news account of the release of the statement, see the *Los Angeles Times,* "Pollutants Can Affect Brain Development, Survey Finds," May 31, 1996.

"Four months later in September 1996 . . ." cites a report titled "Intellectual Impairment in Children Exposed to Polychlorinated Biphenyls in Utero," by Joseph L. Jacobson and Sandra W. Jacobson in the *New England Journal of Medicine,* Vol. 335, No. 11, Sept. 12, 1996, pp. 783–789.

"Another mother-child study in the Netherlands . . ." refers to the study "Effects of Polychlorinated Biphenyl/Dioxin Exposure and Feeding Type on Infants' Mental and Psychomotor Development" by Corine Koopman-Esseboom et al. in *Pediatrics,* Vol. 97, No. 5, May 1996, pp. 700–706.

"Dutch pediatric researchers also conducted . . ." relies on "Immunologic Effects of Background Prenatal and Postnatal Exposure to Dioxins and Polychlorinated Biphenyls in Dutch Infants," by Nynke Weisglas-Kuperus et. al. in *Pediatric Research,* Vol. 38, No. 3, 1995, pp. 404–410.

"Over the past year, two separate assessment efforts . . ." refers to the following publications: "Statement from the Work Session on Chemically Induced Alterations in the Developing Immune System: The Wildlife/Human Connection," in *Environmental Health Perspectives,* Vol. 104, Supplement 4, August 1996, pp. 807–808; and "Pesticides and the Immune System: The Public Health Risks," by Robert Repetto and Sanjay S. Baliga, World Resources Institute, March 1996.

"No aspect of the extensive scientific evidence . . ." introduces a discussion of several new studies examining sperm-count changes in humans and animals. The Scottish study described in the text is "Evidence of deteriorating semen quality in the United Kingdom: birth cohort study in 577 men in Scotland over 11 years," by Stewart Irvine et al. in the *British Medical Journal*, Vol. 312, 24 February 1996, pp. 467–71. The two U.S. studies are (for New York, Minneapolis, and Los Angeles) "Semen analyses in 1,283 men from the United States over a 25-year period: no decline in quality," by Harry Fisch et al. in *Fertility and Sterility*, Vol. 65, No 5, May 1996, pp. 1009–1014, and (for Seattle) "Data from men in greater Seattle area reveals no downward trend in semen quality: further evidence that deterioration of semen quality is not geographically uniform," by C. Alvin Paulsen et al. in *Fertility and Sterility*, Vol. 65, No. 5, May 1996, pp. 1015–1020. The work by Niels Skakkebaek cited in this section can be found in "Evidence for Increasing Incidence of Abnormalities of the Human Testis: A Review," by Aleksander Giwercman, Elisabeth Carlsen, Niels Keiding, and Niels Skakkebaek in *Environmental Health Perspectives*, Vol. 101, Supplement 2, 1993, pp. 65–71. The findings that men volunteering for vasectomies have higher sperm counts than paid sperm donors who had fathered one child was published in "Decline in semen quality among fertile men in Paris during the past 20 years," by Jacques Auger, et al. in the *New England Journal of Medicine*, Vol. 332, No. 5, Feb. 2, 1995, pp. 281–285. The work by Sharpe and Sumpter on the effects of octylphenol and butyl benzyl phthalate—"Gestational and Lactational Exposure of Rats to Xenoestrogens Results in Reduced Testicular Size and Sperm Production." by Richard Sharpe et al.—appeared in *Environmental Health Perspectives*, Vol. 103, No. 12, Dec. 1995, pp. 1136–1143.

"Human health studies are also heightening . . ." cites "Pesticide Appliers, Biocides, and Birth Defects in Rural Minnesota," by Vincent F. Garry et al. in *Environmental Health Perspectives*, Vol. 104, No. 4, April 1996, pp. 394–399.

"Alarming deformities are also showing up" is based on interviews with U.S. Environmental Protection Agency biologist Joseph Tietge of the Mid-Continent Ecology Division of National Health and Environmental Effects Laboratory in Duluth and David Hoppe, a herpetologist at the Morris campus of the University of Minnesota. It also relies on various news accounts in the *Washington Post* (Sept. 30, 1996), Reuters (Oct. 30, 1996), the *Minneapolis Star Tribune* (Sept. 1, 1995, Oct. 23, 1996), and the *St. Louis Post Dispatch* (Oct. 20, 1996).

"Now a new study . . ." cites work by Steven F. Arnold et al. reported in "Differential Interaction of Natural and Synthetic Estrogens with Extracellular Binding Proteins in a Yeast Estrogen Screen," *Steroids* (in press), Jan. 1997.

"With a paper published in . . ." describes "Synergistic Activation of Estrogen Receptor with Combinations of Environmental Chemicals," by Steven F. Arnold et al. in *Science*, Vol. 272, 7 June 1996, pp. 1489–92. For a broader discussion of synergistic effects, see "Synergistic Signals in the Environment," by Steven F. Arnold and John A. McLachlan in *Environmental Health Perspectives*, Vol. 104, No. 10, Oct. 1996, pp. 1020-23, and "Environmental Estrogens," by John A. McLachlan and Steven F. Arnold, *American Scientist*, Vol. 84, Sept.–Oct. 1996, pp. 452–461.

INDEX

Page numbers in **boldface** refer to illustrations.